XI International Road Safety Seminar (GAMBIT 2016)

Road Innovations for Safety

MATEC Web of Conferences Volume 122 (2017)

Gdansk, Poland
17 - 18 November 2016

Editors:

Kazimierz Jamroz
Joanna Zukowska
Aleksandra Romanowska

ISBN: 978-1-5108-4706-4

Printed from e-media with permission by:

Curran Associates, Inc.
57 Morehouse Lane
Red Hook, NY 12571

Some format issues inherent in the e-media version may also appear in this print version.

Printed by Curran Associates, Inc. (2017)

For additional information, please contact EDP Sciences – Web of Conferences at the address below.

EDP Sciences – Web of Conferences
17, Avenue du Hoggar
Parc d'Activité de Courtabœuf
BP 112
F-91944 Les Ulis Cedex A
France

Phone: +33 (0) 1 69 18 75 75
Fax: +33 (0) 1 69 28 84 91

contact-edps@webofconferences.org

Additional copies of this publication are available from:

Curran Associates, Inc.
57 Morehouse Lane
Red Hook, NY 12571 USA
Phone: 845-758-0400
Fax: 845-758-2633
Email: curran@proceedings.com
Web: www.proceedings.com

TABLE OF CONTENTS

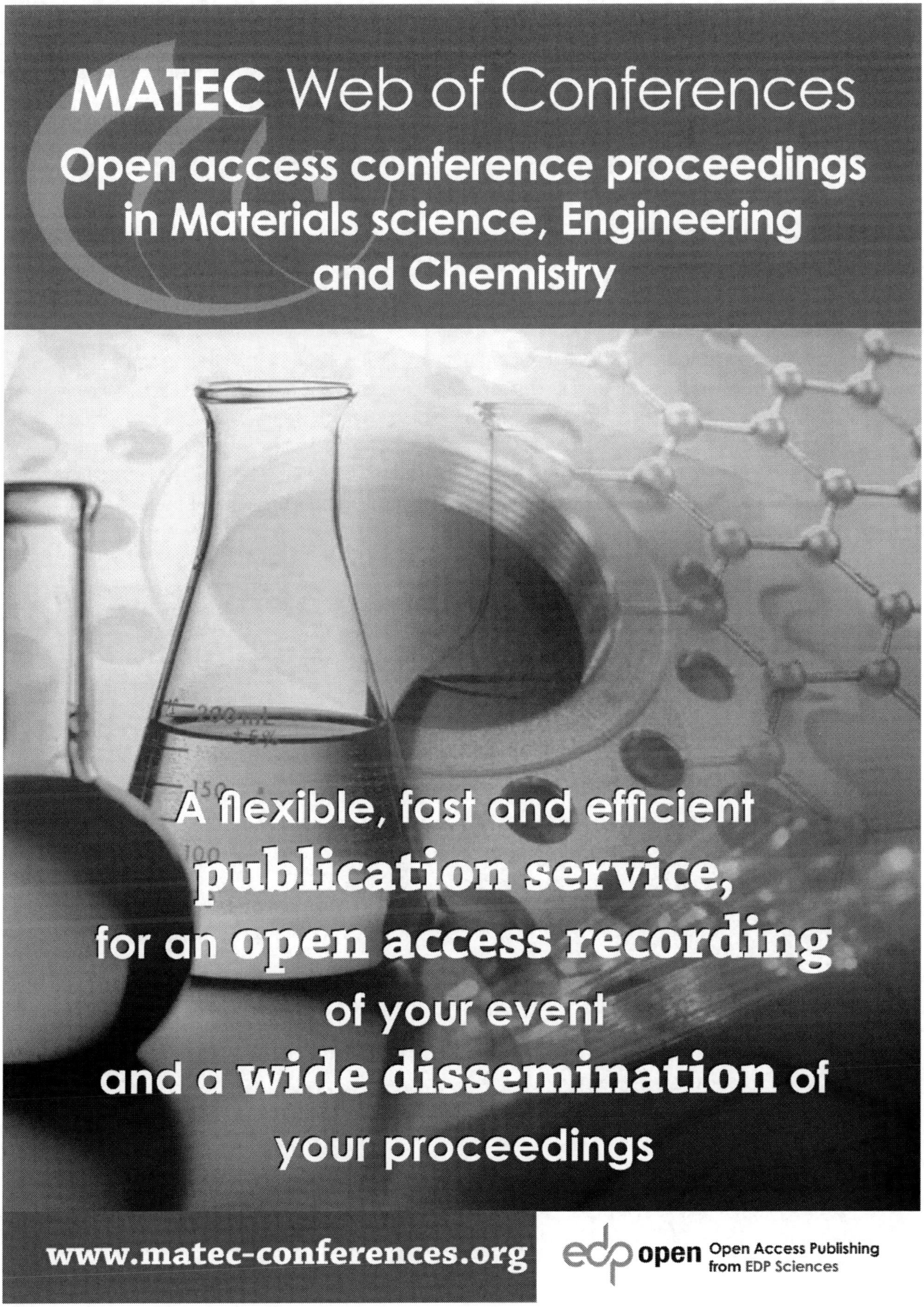
MATEC Web of Conferences
Open access conference proceedings in Materials science, Engineering and Chemistry
A flexible, fast and efficient
publication service,
for an open access recording
of your event
and a wide dissemination of
your proceedings
www.matec-conferences.org
edp open Open Access Publishing from EDP Sciences

XI International Road Safety Seminar GAMBIT 2016

Road Innovations for Safety

Gdansk (Poland), 17-18 November 2016

Editors:

Kazimierz Jamroz
Joanna Żukowska
Aleksandra Romanowska

Organization

Highway and Transport Engineering Department, Faculty of Civil and Environmental Engineering, Gdansk University of Technology

Motor Transport Institute, Poland
Road and Bridge Research Institute, Poland
Foundation of Civil Engineering Development, Gdansk, Poland

Scientific Committee

Kazimierz Jamroz – Chairman
Joanna Żukowska – Scientific Secretary

Ryszard Krystek
Jan Kempa
Mirosław Siergiejczyk
Ghazwan Al-Haji
Zbigniew Kowalewski
Marcin Ślęzak
Jiří Ambros
Ireneusz Kreja
Leszek Smolarek
Ewa Bagińska
Kyamakya Kyandoghere
Adam Tarnowski
Ruth Bergel-Hayat

Francesca La Torre
Piotr Tomczuk
Janusz Bohatkiewicz
Norbert Maliszewski
Marian Tracz
Jacek Chróścielewski
Miroslava Mikusova
Jerry Wekezer
Stanisław Gaca
Marzena Nowakowska
Fred Wegman
Alfredo Garcia
Piotr Olszewski

Krzysztof Wilde
Josef Gnaps
Marek Przyborski
Wojciech Witkowski
Krzysztof Grzelec
Jerzy Pyrchla
Adam Zofka
Péter Holló
Leszek Rafalski
Józef Judycki
Tadeusz Sandecki

Organizational Committee

Kazimierz Jamroz – Chairman
Romanika Okraszewska – Organizational Secretary

Marcin Antoniuk Łukasz
Łazarski
Marcin Budzyński Tomasz
Mackun
Krystian Birr
Lech Michalski
Daniel Bytner
Izabela Oskarbska

Ewa Czarnecka
Jacek Oskarbski
Anna Gobis
Aleksandra Romanowska
Lucyna Gumińska
Maciej Sawicki
Łukasz Jeliński
Barbara Strzebrakowska

Daniel Kaszubowski
Joanna Wachnicka
Wojciech Kustra
Joanna Żukowska
Karol Żarski
Marcin Zawisza

Acknowledgments

The organizers of the GAMBIT 2016 Seminar are grateful for the financial support given by the following institutions:

Foundation of Civil Engineering Development in Gdansk
PrOWERk sp. z o.o., Cracow, Poland
STAR ITS sp. z o.o., Gdansk, Poland
Saferoad sp. z o.o., Wloclawek, Poland

Honorary Patronage

Ministry of Infrastructure and Construction, Poland
National Research and Development Centre, Poland
General Director for National Roads and Motorways
European Road Federation
Gdansk University of Technology

Media patronage

ITS Przegląd | e-droga.pl | Polskie Drogi

edp sciences

MATEC Web of Conferences

Statement of Peer review

In submitting conference proceedings to MATEC Web of Conferences, I certify to the Publisher that I adhere to the **Policy on Publishing Integrity** of the journal in order to safeguard good scientific practice in publishing.

1. All articles have been subjected to peer review administered by the proceedings editors.
2. Reviews have been conducted by expert referees, who have been requested to provide unbiased and constructive comments aimed, whenever possible, at improving the work.
3. Proceedings editors have taken all reasonable steps to ensure the quality of the materials they publish and their decision to accept or reject a paper for publication has been based only on the merits of the work and the relevance to the journal.

Title, date and place of the conference

XI INTERNATIONAL ROAD SAFETY SEMINAR GAMBIT 2016
17-18 NOVEMBER 2016
GDAŃSK

Proceedings editor(s):

PROFESSOR KAZIMIERZ JAMROZ
PROFESSOR JOANNA ŻUKOWSKA

Date and editor's signature

30.06.2017

MATEC Web of Conferences 122, 00001 (2017) DOI: 10.1051/matecconf/201712200001
GAMBIT 2016

Conditions and capacity for implementing Poland's vision zero

Kazimierz Jamroz[1,*], *Wojciech* Kustra[1], and *Joanna* Żukowska[1]

[1]Gdansk University of Technology, Faculty of Civil and Environmental Engineering, Gdansk, Poland

Abstract. Since 1991 Poland has been systematically reducing its traffic hazard. Despite considerable progress Poland is still among the most dangerous countries in the European Union. The key types of actions that help to reduce fatalities include enhancement of pedestrian safety, reducing the number of speeding drivers and eliminating or reducing hazard on the road. The paper presents a brief diagnosis of the state of road safety in Poland and a synthetic evaluation of the implementation of further national road safety programmes. It proposes measures necessary for achieving Vision Zero until 2050 (adopted by the EU) and an assessment of their effectiveness.

1 Introduction

It has been 20 years since the emergence of a new approach to measures to protect road traffic participants from injury and death. Sweden began the implementation of a new road safety policy called Vision Zero in 1996, which was adopted in the Swedish transport plan in 1997 [1,2]. Vision Zero refers to the philosophy of planning actions to improve road safety, although this philosophy was first applied in 1811 in the chemical company DuPont in the USA, where after a tragic accident involving explosives, the basic principles of safety were developed and responsibility for accidents was shifted to management. Moreover, a newly introduced requirement to document and analyse accidents and to create a working environment forgiving employees' mistakes led to enormous improvements in safety. To this day, the DuPont group remains a world leader in the field of occupational health and safety. Findings and ground breaking successes of safety measures applied by DuPont and the activity of Swedish chemical companies in the early 1990 (after an accident in a chemical plant in Seveso and Bohpal) are still important elements of "Vision Zero" which are also applied in the mining industry (mines) and transport [3].

The basic premise of "Vision Zero" in road transport is that serious accidents (fatal or serious injury) are not acceptable, and the long-term goal is to reduce their number to zero. According to the works of its authors [1,2] "Vision Zero" is:

- a philosophy of road safety planning, in which it is assumed that nobody should die or be seriously injured while using the road transport system.
- a long-term business strategy, involving the transformation of the road transport system and a change in the degree of responsibility between road users and vehicles (including the automotive industry) and the road and its surroundings (planners, designers and roads

* Corresponding author: kjamroz@pg.gda.pl

DOI: 10.1051/matecconf/201712200001

maintenance system; the traffic enforcement system, rescue system and health care, government, parliament, etc.).

This approach, which states that nothing is more important than human life, puts road safety before any other assessment criteria of the functioning of the road transport system such as mobility, economy, environment. Ensuring mobility measured by the size and speed of transport is a basic requirement set for the road transport system. Total transport volume (traffic volume, the distance travelled by vehicles) affects the number of accidents, while speed influences the consequences of accidents (the share of fatalities and serious injuries). The search for the solution to the conflict between mobility and safety has stimulated a lot of innovation with increasing investments to develop a safe road transport system. The search goes towards increasing the role of speed surveillance systems or the use of ITS equipment [4]. Thus the question arises whether a non-negotiable ethical approach can compete with the economic approach? Is it possible to achieve the objectives of Vision Zero, how long will it take and at what costs? What social, economic, organisational obstacles can arise during its implementation?

Despite these concerns, many countries have adopted Sweden's idea of improving road safety including Norway, Denmark, Australia, Switzerland, the US [5–8]. Poland also adopted the Vision Zero in 2005 as an ethically legitimate vision of road safety [9].

Despite the growing interest in Vision Zero, there are no guidelines for politicians, transport planners, practitioners in the field of public health, the police and other people implementing road safety programmes according to the rules defined by Vision Zero, particularly at the strategic level. Decisions taken by the political authorities of the country, the Ministry of Transport and the Ministry of Infrastructure, at the strategic (political) level are significant (they have a very broad impact). They are made under conditions of uncertainty or risk arising from limited access to data, high volatility of the phenomenon and the lack of appropriate tools to facilitate decision making. Therefore, a variety of tools to facilitate strategic decision-making is required. To meet these expectations the article presents an analysis of the conditions and possibilities of achieving the objectives of Vision Zero in Poland by 2050 at a strategic level.

2 Implementation of road safety programmes in Poland

The EU is the world leader in road safety. A long-term vision is to eliminate or significantly reduce the number of road fatalities by 2050. Four countries: Germany, France, Italy and Poland produce more than 50% of road accident fatalities in the EU. Beginning from 1991, hazard in road traffic has been steadily decreasing in Poland, in the last 25 years the number of fatalities decreased by over 60%. Despite such significant progress, Poland is still among the most dangerous countries in the European Union. Many times in the past 15 years, Poland has been in the first (worst) place among the countries of the European Union, taking into account the demographic mortality rate in road accidents. In 2015 Poland was number five in a ranking of 28 countries. Are things really that bad?

The risk of being a fatal victim in relation to the number of inhabitants in Poland is twice as high as the EU average, and even three times higher than the UK, the Netherlands and Sweden. But Poland is learning from the experience of other countries and uses it in a positive way (Fig. 1).

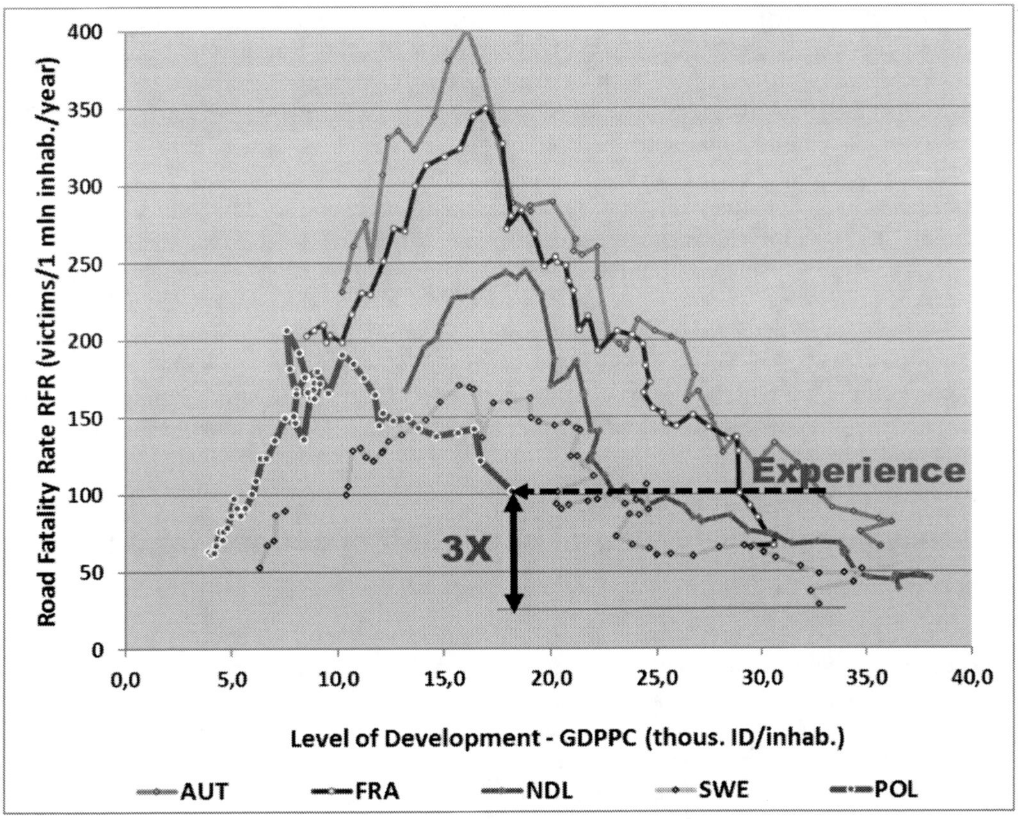

Fig. 1. Comparison of changes in the RFR in Poland with selected countries.

25 years ago (1991) in Poland the upward trend in the number of fatalities was broken. In 2005 Vision Zero was adapted as Poland's vision of road safety. Poland benefits from the experience of other countries, learns from them and uses the knowledge in its efforts to improve safety. This can be seen in Figure 1 with a comparison of road safety levels in Poland and France. The two countries have reached a level of safety – the RFR = 100 fatalities per 1 million inhabitants at different periods of time and at different level of socio-economic development (measured by the level of GDP per capita in international dollars). In 2003 France had a GDP per capita = 30 thousand ID, in Poland in 2012 the GDP per capita = 19 thousand ID, i.e. a much lower level of socio-economic development. This demonstrates the early lessons learned, the result of taking advantage of the experience of advanced countries (the Netherlands, Sweden, the United Kingdom, France, etc.) in an effort to reduce the road accident mortality rate and to take systematic and programmatic actions.

Progress in the application of improvements and actions for road safety is slower than in other countries of the EU. The national road safety programmes have been the basis for a fatality reduction. The National Road Safety Programmes included: GAMBIT 1996, GAMBIT 2000, GAMBIT 2005 [10] and National Road Safety Programme by 2020 [11], GAMBIT for National Roads (2007-2013), as well as numerous regional and local road safety programmes [12]. In spite of the advanced implementation of the programmes the main problems of road safety in Poland persist: dangerous behaviour of road users (driving at dangerous speeds: 50% of drivers exceed speed limits, 27% of all fatalities); lack of pedestrian protection (pedestrians – 35% of deaths in road accidents, 1st place in the EU); a poorly organised system of road safety (victims who died within 30 days after the accident still constitute 32% of total deaths); poor quality of road infrastructure (34% of deaths on national roads which carry more than 25% of the traffic load); an unforgiving road environment (trees, poles) is the cause of 18% of total deaths. But there are positive effects of the road safety programmes, such as drink driving campaigns which led to a significant reduction in the share of fatal accidents of this type from 24% in 1995 to 11% in 2014 (which gives 25th place in the ranking of the EU).

Therefore, Poland's road safety programmes (implemented in the last 20 years) put the emphasis on strengthening pedestrian safety, reducing excessive speeding and eliminating or reducing the impact of the various sources of danger (hazard) on the road. The progress of the programmes was analysed and assessed and shows that further measures to reduce the risk of traffic in Poland should be designed in accordance with the concept of the "three D's" on: continuation of the actions on infrastructure, including the completion of the network of higher class roads (motorways and express roads), development of safety management systems (with particular emphasis on automated supervision of dangerous behaviour of road users) and the promotion of safety culture among politicians, decision makers and road traffic participants.

3 Analysis and evaluation of scenarios of actions for achieving Vision Zero

In trying to answer the previous questions, several scenarios of the country's demographic and economic development and road safety system development were prepared. Using the previous studies [13–15] four scenarios were analysed that differ in terms of: population, national product and actions for road safety.

Scenario S1 represents low socio-economic development (18 thousand ID per capita in 2014 to 50.5 thousand ID per capita in 2050) and slight decrease in the number of inhabitants from 38.5 in 2014 to 36.3 million in 2050, lack of additional activities and interventions as well as limited operations of automated speed supervision systems (including reducing the number of speed cameras).

Scenario S2 represents low socio-economic development (50.5 thousand ID per capita in 2050) with an average decrease in the number of inhabitants to 33 million in 2050, initially a low number of additional actions, low number of actions and with time increase in the number of actions.

Scenario S3 represents average socio-economic development (62 thousand ID per capita in 2050) and an average decrease in the number of inhabitants to 33 million in 2050, quite a large number of additional organizational and infrastructural actions and change in the behaviour of road users.

Scenario S4 represents average socio-economic development (74 thousand ID per capita in 2050) and a high decrease in the number of inhabitants to 29.7 million in 2050, a large number of additional systemic actions (development of the health system, road safety system, automated traffic surveillance systems (speed cameras), decrease in the mobility of citizens, the development of a safe road infrastructure network (construction of motorways and expressways), etc.

To evaluate the scenarios the number of fatalities F was adopted. To estimate the number of road fatalities in the future from projected or assumed changes in social, economic and transport systems, the author's forecasting model was used, developed in the framework of the following works [12–16]. To estimate the number of fatalities, factor models were adopted in the general form shown in equations (1), (2), (3):

$$F = P \cdot RFR_b \cdot MF_c \tag{1}$$

where:

$$RFR_b = \beta_o \cdot GDPPC^{\beta_1} \cdot VTKPC^{\beta_2} \cdot exp(-\beta_3 \cdot GDPPC - \beta_4 \cdot LEI - \beta_5 \cdot CPI + \beta_6 \cdot ACPC + \beta_7 \cdot DPR - \beta_8 \cdot DME - \beta_9) \quad (2)$$

$$MF_c = \gamma_o \cdot exp(-\gamma_1 \cdot LEI - \gamma_2 \cdot DME - \gamma_3 \cdot \ln(FV) + \gamma_4) \quad (3)$$

where:
F – number of fatalities on the country's road network (fatalities/country/year), *P* – population (M inhabitants/country/year), RFR_b – base road fatality rate (fatalities/1M inhabitants/year), MF_c - country modification factor, *GDPPC* – gross national product per capita (thousands ID/inhabitant/year), (PPP, constant 2005, international $), *VKTPC* – average vehicle kilometres travelled per year per capita (km/population/year), *LEI* – life expectancy index, *CPI* – corruption perception index, *ACPC* – alcohol consumption per capita (l/inhabitant/year), *DPR* – density of paved roads dependent on demography (km/1 M population), *DME* – density of motorways and expressways dependent on demography (km/1 M population), *FV* – number of speed cameras in use (cam./year), β_i, γ_i – equation coefficients.

The base model – road fatality rate (*RFR*) was developed from empirical data collected for selected countries which have trusted statistics of road fatalities, and in which there were no significant disturbances in the course of changes in accident mortality. Data was collected on the number of fatalities and the number of parameters characterising the selected countries in different years i.e. variables: geographic, demographic, economic, social, automotive, road and transport. The analyses helped to establish that in the base model the most important factors influencing the change of the *RFR* indicator are: the level of socio-economic development represented by the individual gross domestic product (*GDPPC*), the health care system and societal care for health represented by the index of life expectancy (*LEI*), inhabitants mobility level, measured by the average distance travelled by vehicles (*VKTPC*) and the structure of the road network represented by density of motorways and expressways (*DME*).

The application of the MF_c coefficient modified the base model to match the national data. The result was a decrease in the mean squared error of the match of the estimated value F to the actual data. This allowed the model to be applied to the method for forecasting long-term changes in Poland's road fatalities. The resulting model correction (3) shows that Poland's road deaths can be reduced additionally through: *LEI* level of health care, development of the motorway and expressway network *DME* and development of an automated speed surveillance system *FV*.

Using the models for different scenarios, forecasts of changes were developed in the individual independent variables, used in the formulas (2) and (3). Next the number of deaths in the period 2014-2050 was estimated for the adopted scenarios. The results indicate that the expected further socio-economic development in Poland and the resulting changes in other independent variables (according to the trends currently observed among the countries with the highest road safety) will bring a systematic reduction in road fatalities. However, as the presented scenarios show, these changes will be insufficient to achieve the Vision Zero of road fatalities in 2050.

If actions are taken as envisaged in scenario 3 or 4, the result can be up to 70,000 lives saved which means that Vision Zero can be achieved. Lack of consistency (e.g. removal of speed surveillance systems) can be dangerous as has been shown in scenarios 1 and 2. In fact, the risk of fatal accidents may rise in the initial period and there may be a significant reduction in the number of lives saved. This means that in these cases the implementation of Vision Zero will be very difficult, or even impossible.

 DOI: 10.1051/matecconf/201712200001

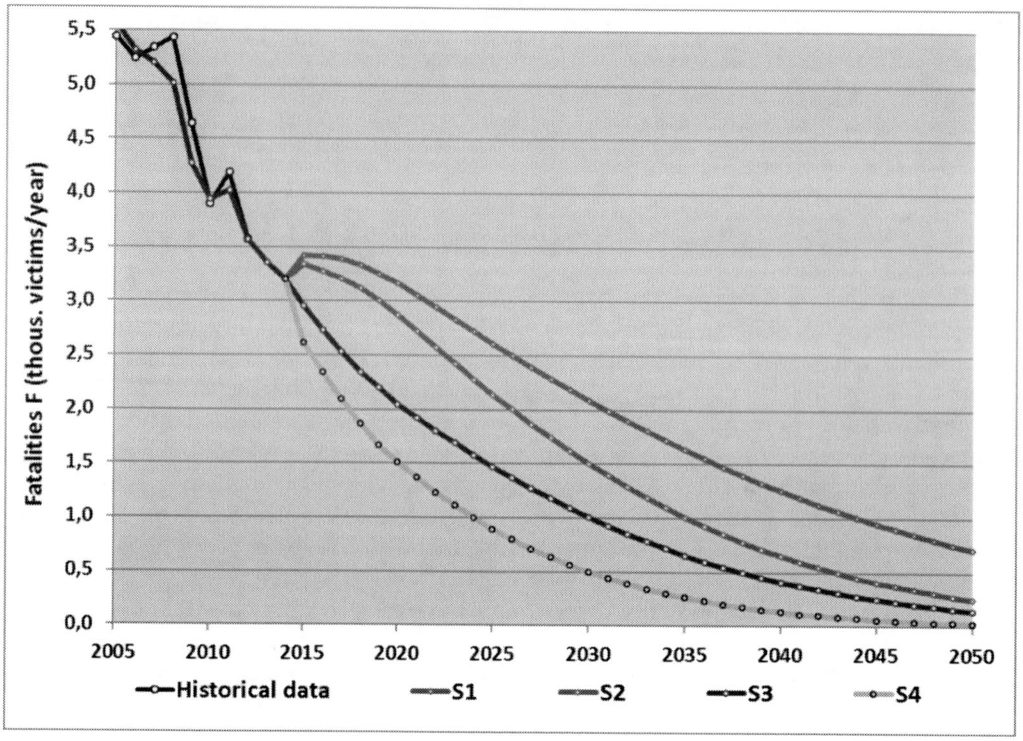

Fig. 2. Forecast of fatalities in Poland by 2050, taking into account different scenarios of external factors.

4 Conclusions

The results of the research and analyses made it possible to reach the following conclusions:

1. The trend change in the number of deaths in Poland has been favourable, since 1991 we have observed a systematic decrease in the "tragic harvest" on Polish roads.
2. Our society still pays too high a price (increase in fatalities and serious injuries) for unproven, often irresponsible and populist actions of the authorities (e.g. frequent changes in the structure and scope of the speed surveillance system, lack of decisive action for the protection of pedestrians, a road network with low standards of road safety).
3. The models for estimating the number of fatalities, proposed in this paper, help to estimate the risk of introducing selected road safety measures at the strategic level.
4. The analyses confirm that it is possible to achieve the objectives of Vision Zero in Poland around 2050, but only under the condition that numerous actions are carried out, with particular emphasis on: pedestrian protection, development of a speed management system, construction of safe roads, change in the safety culture, etc.
5. It is necessary to conduct further research (e.g. on the number of victims seriously-injured, accident costs, etc.) to improve the presented models and further develop the safety measures; the models should be used to develop and implement in practice modern management methods of road safety.

References

1. C. Tingvall, The Swedish "Vision Zero" and how parliamentary approval was obtained, in: Road Saf. Res. Policing, Educ. Conf., (Wellington, New Zealand, 1998)
2. C. Tingvall, N. Haworth, Vision Zero - An ethical approach to safety and mobility, in: 6th ITE Int. Conf. Road Saf. Traffic Enforc. Beyond 2000, (Melbourne, 1999)
3. H. Ehnes, Vision Zero - The New Global Strategy for Safe Mining, in: Int. Symp. Cult. Prev. - Futur. Approaches From Risks To Vis. Zero, (Finnish Institute of Occupational Health, 2014)
4. L.C. Wadhwa, Vision Zero requires five-star road safety system, in: Road Saf. Res. Polic. Educ. Conf., (Melbourne, Australia, 2001)
5. A. Erke, R. Elvik, Making Vision Zero real: Preventing pedestrian accidents and making them less severe. TOI report 889/2007 (Oslo, 2007)
6. D. Atkins, M. Granhed, Vision Zero: Applying road safety to avalanche safety, in: Int. Snow Sci. Work., (Anchorage, Alaska, 2012)
7. A. Fleisher, M.L. Wier, M. Hunter, A Vision for Transportation Safety: 1 A Framework for Identifying Best Practice Strategies to Advance Vision Zero, TRB. (2016)
8. WHO, Global Plan for the Decade of Action for Road Safety 2011-2020 (2011)
9. K. Jamroz, L. Michalski, S. Gaca, Road safety programmes as an effective tool for developing system-based road safety policies, J. Konbin. **1** pp. 155–160 (2006)
10. L. Michalski, K. Jamroz, J. Żukowska, UN Decade of action for road safety in the national road safety strategy until 2020 – Polish approach, in: 16th Int. Conf. Road Saf. Four Cont., (Beijing, China, 2013)
11. National Road Safety Programme for 2013-2020 (Polish National Road Safety Council, Warsaw, 2013)
12. K. Jamroz, Is Vision Zero realistic in Poland?, in: 4th Int. Conf. Road Saf. Simul., (Roma, 2013)
13. K. Jamroz, The impact of road network structure and mobility on the national traffic fatality rate, Procedia - Soc. Behav. Sci. **54** pp. 1370–1377 (2012). doi:10.1016/j.sbspro.2012.09.851
14. K. Jamroz, L. Smolarek, Road safety management tools for country strategic level, in: 16th Road Saf. Four Cont. Conf., (Beijing, China, 2013)
15. K. Jamroz, W. Kustra, M. Budzyński, J. Żukowska, Pedestrian Protection, Speed Enforcement and Road Network Structure the Key Action for Implementing Poland's Vision Zero, Transp. Res. Procedia. **14** pp. 3905–3914 (2016). doi:10.1016/j.trpro.2016.05.479
16. K. Jamroz, *Method of risk management in highway engineering. Metoda zarządzania ryzykiem w inżynierii drogowej* (Gdansk University of Technology, Gdansk, 2011)

Study method for pedestrian behaviour in the area of pedestrian crossings located at tram stops

Marek Bauer [1,*], and *Wiesław* Dźwigoń [1]

[1]Cracow University of Technology, Faculty of Civil Engineering, Cracow, Poland

Abstract. The paper presents the problem of pedestrian safety at stops, with stops designed for opposite directions. Stop areas with a predominantly pedestrian traffic, where pedestrians cross tram tracks, are not protected by traffic signals. The article presents a method for conducting a study to identify types of pedestrian and tram driver behaviour, developed on the basis of a pilot study. This method helps to evaluate, inter alia, the tendency of pedestrians to risk going across the tram track before an oncoming tram, depending on how far it is from the pedestrian crossing. It also helps to determine the probability of the tram slowing down or emergency braking when crossing the pedestrian crossing, and the probability of difficulties with the departure of the tram, caused by pedestrians.

1 Introduction

Nowadays, trends can be seen in European cities aiming to increase the role of walking and public transport [1]. This is especially true of downtown areas, where excessive car traffic degrades urban spaces and deteriorates living conditions of the citizens [2]. The main reason for this is the considerable and often increasing number of journeys within the downtown area which cannot be satisfied with car transport. The seemingly favourable development of new roads causes even more congestion and takes away space from pedestrians who should be given special priority in the city centre [3]. Therefore, the gradual shift from the previous policy enabling a free use of the car in city centres is a highly desirable phenomenon, especially for medium-sized and large cities [4]. The scope of possible actions is relatively broad and includes solutions facilitating public transport traffic, supporting walking and cycling and, an area which still remains unaccepted by some users of the transportation system, restricting the car [5]. Only if used skilfully, can these solutions create a chance for success and keep the area attractive for the public and visitors while ensuring good access to the area. The simultaneous use of solutions favouring vulnerable road users and public transport vehicles is only seemingly an easy task. Above all, pedestrians expect unrestrained access. Passengers of public transport expect a quick connection to the downtown area which must be attractive, i.e. relatively quick, particularly when it comes to the tram network which is more susceptible to traffic disturbances [6]. Ensuring the right balance between the degree of enhancement of pedestrian traffic and public transport vehicles has considerable importance. The most important problems observed in city centres are conflicts between pedestrians and trams within tram stops, particularly when the stops are located near busy

* Corresponding author: mbauer@pk.edu.pl

footways with pedestrian streams crossing stops when the trams are arriving even if they do not want to use the tram. The key role in this sense is played by the diversity of pedestrian behaviours and variable awareness of the traffic and braking capabilities of the trams.

The present study discusses the issues related to the comfort and safety of using transport infrastructure in the potential conflict zone between pedestrian and tram traffic. The original proposal for the study methodology includes road user behaviours (pedestrians and tram drivers) in the vicinity of pedestrian crossings located near tram stops. Conflicting interests of vulnerable road users (safety and comfort) and passengers of public transport (comfort and speed) must be reconciled in these areas.

2 Current approach to the study of conflicting situations between pedestrians and trams

The number of accidents and collisions that took place on individual tram stops is usually not relevant, as the accidents take place relatively rarely [7]. Moreover, studies of hazardous situations enable a description of atypical or temporary solutions - which should be tested prior their eventual dissemination. On the basis of the definition accepted in 1997 in Oslo, it can be stated that "pre-conflicting road user behaviour is a situation, when the users violate the established thresholds for road behaviour, yet which do not result in a road accident, and which do not force a change in the behaviour of other road users in order to avoid the accident" and "traffic conflict is a situation, when at least one of the users takes action to prevent an accident". The basic indexes evaluating traffic conflicts include:

- CS (Conflicting Speed) - this is the speed of road user, who began evasive actions (just before starting the action) [8];
- TA (Time to Accident) - time which passes from the taking of an evasive action by at least one road user to the moment of a potential accident, if both road users continued movement at the same speed and the same direction;
- TTC (Time To Collision) - time to the collision of two objects, if their trajectories and speeds remain unaltered. The lower value of TTC means a higher likelihood of the accident with more serious consequences [9].

However, the above indexes are more appropriate for the study of "vehicle-vehicle" conflicts than for "vehicle-pedestrian" conflicts [10], as they concern independent vehicles already in motion. Whereas in the case of "vehicle - pedestrian" conflict, particularly within stops, the conflicting situations often stem from the incorrect evaluation of the possibility to enter the track (pedestrian) or incorrect evaluation of the possibility to leave the stop (tram driver).

Therefore, techniques for the study of road user behaviour have been introduced and enhanced. They are typically conducted in a traditional form, with observers detecting conflicting situations and entering the types and numbers to the questionnaires, often as supplementary research to pedestrian and vehicle traffic measurements. Currently, a video technique is increasingly commonly used for this purpose enabling a lower number of participants to the observations and offering better accuracy of the conducted study. However, independently of the technical feasibilities of research, selection of the observed behaviours and correct, unambiguous definition of conflicting situations remain the most important issue. It largely depends on the approach of the researchers. A major step in the development of traffic conflicts was the research conducted within the project CiViTAS CARAVEL [11], implemented in order to enhance the attractiveness, comfort and safety of city transport. In the study, conflict measurements included three types of stops: with passenger exchange in the lane, with elevated lane (so called Vienna-like tram stop) and with elevated lane and island platform. The study did not analyse opposite stops located in areas with shared tram-pedestrian function.

3 Proposed method for the study of conflict situations between pedestrians and trams within the area of opposite tram stops

Conflicting situations between pedestrians and trams within tram stops typically take place while pedestrians are crossing tracks - and the proposed study will be most relevant in this field. Moreover, only crossings without traffic lights were included. In the proposed method, the level of safety is assessed using the analysis of frequency and potential outcomes of hazardous situations (traffic conflicts). For obvious reasons, the method must comply with the current legislation including the rules on priority. These rules are discussed in Article 13 of the Road Traffic Act [12]; they demonstrate the obligation to cross the track separated from the traffic lane only in the designated area, and where a passenger island situated on a public transport stop is connected to a pedestrian crossing, reaching and leaving the stop is allowed solely on the crossing. However, the most important issue is that the pedestrian is obliged to take special care prior to entering the traffic lane. While on the crossing, the pedestrian has priority over all vehicles, including trams. However, the pedestrian does not have priority over a tram while waiting to enter the pedestrian crossing, which is regulated by Article 14 of the above Act, prohibiting (among others) stepping onto the lane directly before an approaching vehicle, including pedestrian crossings. An issue to be addressed is evaluating the safe distance from the tram while entering the pedestrian crossing. The pedestrian is not allowed to step onto the tracks before an approaching tram, whereas the tram driver approaching a pedestrian crossing must maintain special attention and travel at a speed enabling him to stop before the crossing if there is a pedestrian on the crossing.

The proposed method includes the general behaviour of pedestrians who cross tracks as well as the behaviour of tram drivers. In principle, the method is to be used for determining the use of infrastructure dedicated to pedestrian traffic, as well as for determining the frequency of traffic conflicts to decrease their prevalence. It is also intended to help with determining the causes of hazardous behaviour of pedestrians and tram drivers. In the proposed method, pedestrian traffic volumes are recorded on the entire width of pedestrian crossings and outside them. Pedestrian traffic volumes are recorded for both possible traffic directions and four possible tram positions: empty tram tracks - no approaching trams (on both sides of pedestrian crossings), tram arriving at a stop (stopping before the platform), tram leaving a stop after the passenger exchange is concluded and simultaneously: tram approaching the stop and tram leaving the stop (opposite direction).

If crossing tram tracks is physically possible on both sides of the tram stop complex, traffic volumes must be recorded for both sides, even in the case when no dedicated pedestrian crossing is designated to one of the sides. The information on the percentage of people crossing outside the pedestrian crossing is part of a study of irregular events. When tracks are crossed at the approach of tramways nearing the stop, certain pre-conflicting and conflicting situations occur, therefore distances from the potential point of conflict were included. This is important because the results help to identify the tendencies of pedestrians to take the risk of crossing tracks when the tramway is nearing the potential conflict point. The following types of behaviours of pedestrians crossing tracks at the approach of a tram were assumed:

- a pedestrian crosses the tracks when there is no tram or the tram is at least 20 meters away, which means that it can stop safely and comfortably (for the passengers) travelling at a speed not exceeding 30 km/h - in practice, this means a correct and safe crossing before an approaching tram (Figure 1);

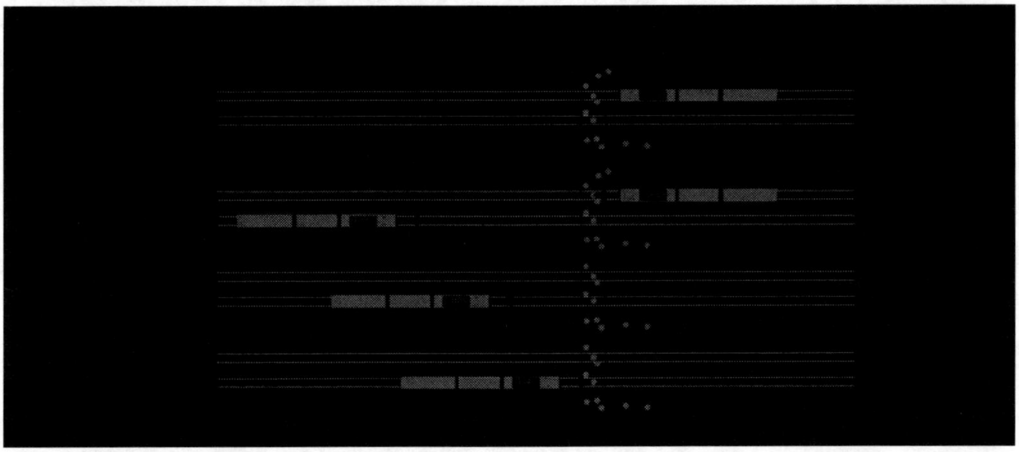

Fig. 1. Diagram of traffic conflicts between pedestrians and trams.

- a pedestrian crosses the tracks when the tram is at a distance of 10–20m, which corresponds to the possibility for a safe stop of the tram nearing the platform at the speed of up to approx. 16 km/h;
- a pedestrian steps on the tracks when the tram is at a distance lower than 10m, which is linked to a considerable risk of a pedestrian being hit, even at a relatively low speed;
- running onto the track when the tram is at a distance lower than 10m, which may result in the tram driver not taking any action (surprised by the appearance of the pedestrian).

On the other hand, in the case of recording pedestrian traffic volumes on a pedestrian crossing situated at the front of the stop with ongoing tram-passenger exchange, we have proposed the inclusion of three types of pedestrian behaviours:

- pedestrian entering the track when the tram remains at the stop;
- entering the track when the tram drives off from the stop - which results in a sound signal, possible deceleration at the start or even emergency stop of the tram;
- running onto the track when the tram accelerates from the stop, which is particularly dangerous due to the possibility of remaining unnoticed by the tram driver.

The following behaviour types were taken into consideration for tram drivers:

- free tram passage – no pedestrians;
- unproblematic start from the stop across the pedestrian crossing, due to the absence of pedestrians on and at the entrance to the crossing;
- passage across the crossing when pedestrians are waiting to step onto the crossing, but they are not on it - treated as correct passage across the crossing, often accompanied by sound signalling, tram deceleration or stop occurring sporadically;
- passage across the crossing when pedestrians are on the crossing, which is almost always accompanied by speed reduction or emergency stop;
- accelerating from the tram stop when pedestrians are on the crossing - depending on the situation, linked to a sound signal, deceleration or emergency stop.

The recorded behaviours of pedestrians and tram drivers must be accompanied by assessment indicators. A common approach is proposed, based on stratum weights. Determination of the share of individual behaviours described above and their reference to pedestrian and tram traffic volumes is possible.

4 Example of the method in practice: study of conflicting situations within the tram stop "Dworzec Główny Zachód" in Krakow

The study of conflicting situations was performed following the presented methodology in several stop areas in Krakow. Figure 2 shows an example of the shares of passengers crossing the track in one of the analysed pedestrian crossings in relation to approaching trams.

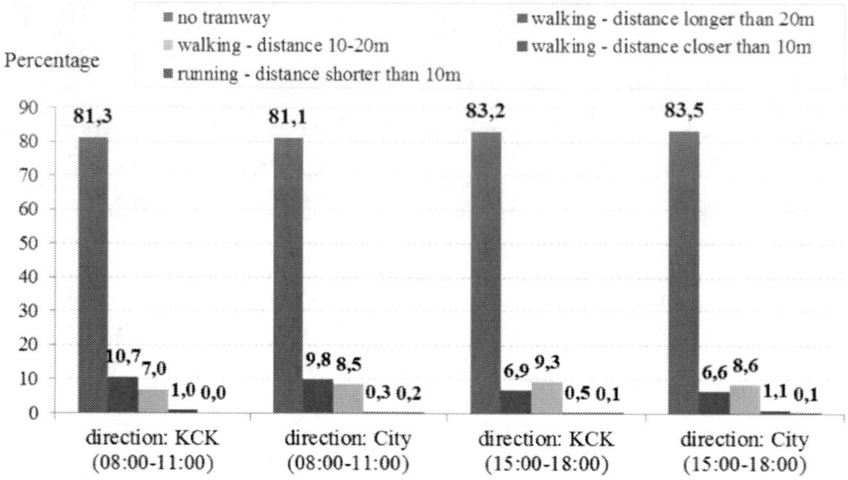

Fig. 2. Share of individual pedestrian behaviours on the analysed pedestrian crossing.

As can be seen, about 1% of pedestrians behave very riskily, in addition, to nearly 8% trying to pass when the tram is 10-20 m from the pedestrian crossing, which can also cause conflicts. And this "distance dilemma" will be particularly closely studied in the future. In this case, the tram track is located in the pedestrian area. Despite being separated, pedestrian and cycling traffic does occur there sporadically. It should be noted, however, that in reality almost the entire space between the above pedestrian crossings does not constitute a spatial barrier - and it is used by pedestrians to cross the tracks. It can also be noted that the share of passages outside of designated places is significant, amounting to 20% and that it varies depending on the direction and time of day. The drivers respond by using sound signals (7-16%), decelerating significantly (4-11%) or even emergency stopping (0-4%). This means that the drivers are prepared for unexpected behaviour of pedestrians and drive at safe speeds.

5 Conclusions

Measurement of hazardous situations occurring at stops and crossings requires further discussion. At this stage of analysis, other features were not investigated such as weather conditions, pedestrian age, visibility and parameters of pedestrian crossings. The absence of such wider studies means that the presented methodology and results should be considered as a pilot. It is important to define the safety and hazardous situations. Pedestrian behaviour is influenced by the location of sources and destinations of traffic and footways. The presented method can be – with some adaptation - used for the study of conflicting situations between trams and bicycles.

 DOI: 10.1051/matecconf/201712201001

References

1. P. Nijkamp, A.H. Perrel, Sustainable cities in Europe (Routledge, 2014)
2. K. Nosal, W. Starowicz, Evaluation of influence of mobility management instruments implemented in separated areas of the city on the changes in modal split, Arch. Transp. **35** pp. 41–52 (2015). doi:10.5604/08669546.1185186
3. K. Jamroz, W. Kustra, M. Budzyński, J. Żukowska, Pedestrian Protection, Speed Enforcement and Road Network Structure the key Action for Implementing Poland's Vision Zero, Transp. Res. Procedia. **14** pp. 3905–3914 (2016). doi:10.1016/j.trpro.2016.05.479
4. D.S. Vale, Transit-oriented development, integration of land use and transport, and pedestrian accessibility: Combining node-place model with pedestrian shed ratio to evaluate and classify station areas in Lisbon, J. Transp. Geogr. **45** pp. 70–80 (2015). doi:10.1016/j.jtrangeo.2015.04.009
5. A. Spickermann, V. Grienitz, H.A. von der Gracht, Heading towards a multimodal city of the future?, Technol. Forecast. Soc. Change. **89** pp. 201–221 (2014). doi:10.1016/j.techfore.2013.08.036
6. D. Esztergár-Kiss, C. Csiszár, Evaluation of Multimodal Journey Planners and Definition of Service Levels, Int. J. Intell. Transp. Syst. Res. **13** pp. 154–165 (2015). doi:10.1007/s13177-014-0093-0
7. L. Michalski, K. Jamroz, Technika konfliktów ruchowych i możliwości jej zastosowania, Transp. Miej. **8** pp. 149–152 (1989)
8. C. Hydén, The development of a method for traffic safety evaluation: the Swedish traffic conflict technique, Doctoral Thesis, Lund University, 1987
9. J.H. Kraay, A. van der Horst, Trautenfels – study: a diagnosis of road safety using the Dutch Conflict observation technique, DOCTOR (Report R–85–53) (Institute for Road Safety Research SWOV, Leidschendam, 1985)
10. S. Gaca, M. Kieć, A. Zielinkiewicz, Analyses of headways for assessment of risk rear-end collision, in: 3rd Int. Conf. Road Saf. Simul., (Indianapolis, 2011)
11. A. Rudnicki, Innovations for sustainable urban transport: experiences from accomplishment of EU Civitas Caravel project (Cracow University of Technology, 2010)
12. Act of 20 June 1997 - Road Traffic Act, OJ 1997 No. 98 Pos. 602 (Poland, 1997)

Pedestrian safety in road traffic – studies, recommendations and proposed improvements

Marcin Budzyński[1,*], *Lucyna* Gumińska[1], *Łukasz* Jeliński[1], *and Mariusz* Kieć[2]

[1]Gdansk University of Technology, Faculty of Civil and Environmental Engineering, Gdansk, Poland
[2]Cracow University of Technology, Faculty of Civil Engineering, Cracow, Poland

Abstract. Pedestrians are involved and most frequently the victims of every third road accident in Poland. Pedestrian accidents most frequently occur in complex circumstances, as a result of many factors related to the behaviour of drivers and pedestrians. The basic parameters that determine road safety include the perception of traffic and visibility on the road. The paper will present the results of research conducted within the project commissioned by the National Road Safety Council Secretariat titled "The methodology of systematic study on pedestrian behaviour and pedestrian - car driver relations". The authors will present the impact of location, type of cross-section and other selected parameters on the behaviour of drivers and pedestrians in the area of pedestrian crossings. The paper will also present recommendations for the design of pedestrian crossings and monitoring road user behaviour at pedestrian crossings.

1 Introduction

Every third road accident in Poland involves a pedestrian as a participant or, most of the time, a casualty. Pedestrian accidents are usually the result of complex situations and the outcome of a number of factors related to driver and pedestrian behaviour and road infrastructure. Safety depends largely on how well the traffic condition is perceived and on visibility in traffic. The paper presents the results of analyses of methodologies for systematic studies of pedestrian behaviour and pedestrian-driver relations (for the Secretariat of the National Road Safety Council) [1]. The effects of the location of the site, type of cross-section and other selected parameters on pedestrian and driver behaviour are demonstrated. Recommendations for pedestrian crossing design are also provided.

2 Method for systematic studies of pedestrian-driver behaviour

2.1 Study description

The main objective of the work was to develop a methodology for systematic studies of pedestrian behaviour and pedestrian-driver relations [2–11]. The aim is to make walking more attractive and improve pedestrian safety. The work involved tests on site and surveys in 69 test points across the regions of Pomorskie and Małopolskie. The areas vary in terms of the development, road cross-section and location of pedestrian crossings (Fig. 1).

* Corresponding author: mbudz@pg.edu.pl

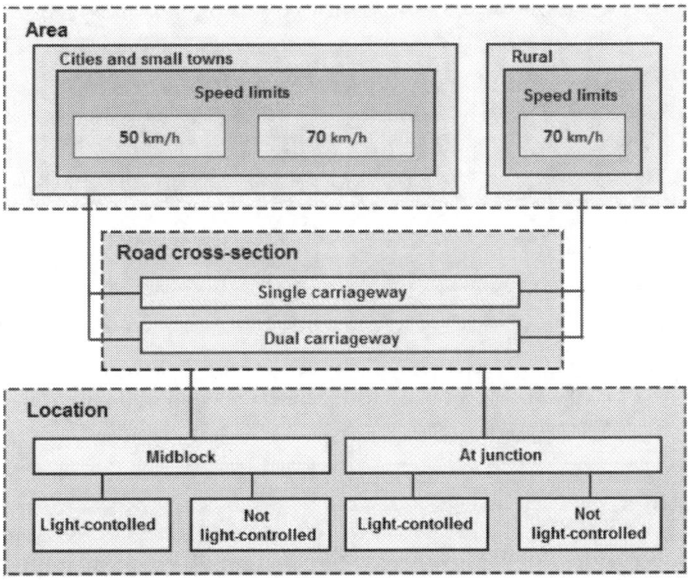

Fig. 1. Diagram of test points selection [1].

One of the site parameters was the speed of vehicles approaching the pedestrian crossing. A set of digital cameras were used for top quality recording. The location of the cameras made sure that they covered the entire crossing as well as the approach (app. 100 m.) with test point cross-sections spaced every 10 m. They were then used to elaborate vehicle speed profiles in the following situations: pedestrian waiting to cross the road, pedestrian crossing the road and no pedestrians on or around the crossing. In addition, vehicles in free-flow and selected pedestrian and driver behaviours were taken into account.

As well as studying pedestrian behaviour, the project also looked at the distance between the pedestrian and road as they wait to move onto the crossing, whether the pedestrian checked the traffic situation before or during crossing the road and pedestrian age structure and sex. In addition, a survey was conducted among pedestrians after they had crossed the road and among drivers who were parked close to the test points. The objective was to identify the main problems as perceived by pedestrians and drivers and identify driver behaviour on pedestrian crossings based on the situations depicted below (Fig. 2).

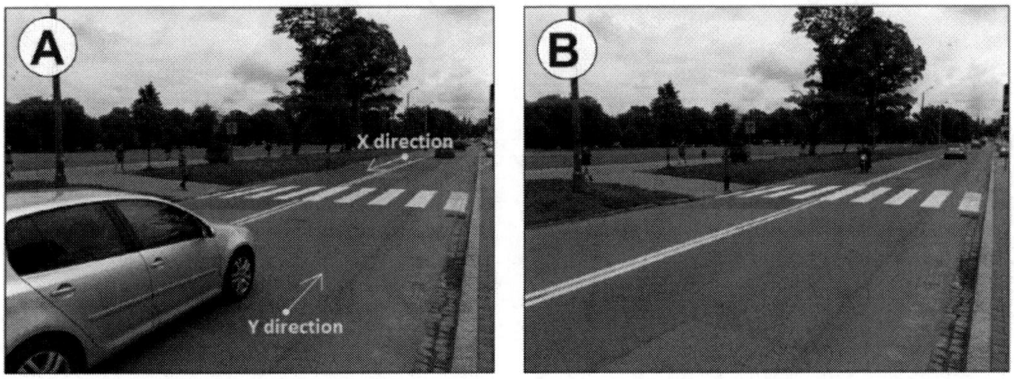

Fig. 2. Situations presented to surveyed drivers [1]: A) pedestrian approaching the crossing, B) pedestrian waiting before the crossing, C) pedestrian moving onto the crossing, D) pedestrian on the crossing.

2.2 Study results

The data collected (video footage) were used to analyse driver behaviour. It was found that when traffic is light-controlled, 3% of drivers do not stop for the red light as required. In 50 km/h areas (test points in towns and villages) app. 40% of drivers exceed the speed limit (Fig. 3a) and in non-built-up areas with a speed limit of 70 km/h, app. 30% of drivers go over the prescribed speed limit (Fig. 3b), a clear indication of the need to apply speed management solutions and enforcement.

In towns and villages for all types of cross-sections, vehicle speeds are lower if pedestrians are waiting to cross as opposed to where there are none (by 16 km/h for 1x2 lanes cross-section, by 16 km/h for 1x4 lanes, by 2 km/h for 2x2 lanes and 2x3 lanes). The lowest speed on approaching a crossing with no pedestrians waiting, at 10 m from the crossing was recorded for the 1x2 lanes cross-section with a refuge island (30 km/h less than for 1x4 lanes, 23 km/h less than for dual carriageways (Fig.4). The lowest speed on approaching a crossing with pedestrians waiting was recoded for 1x2 lanes cross-section with no refuge (16 km/h less than for 1x4 lanes, 22 km/h less than for dual carriageways). The lowest speed on approaching a crossing being crossed by a pedestrian was recorded for 1x2 lanes with a refuge but the differences between the cross-sections are quite small.

a)

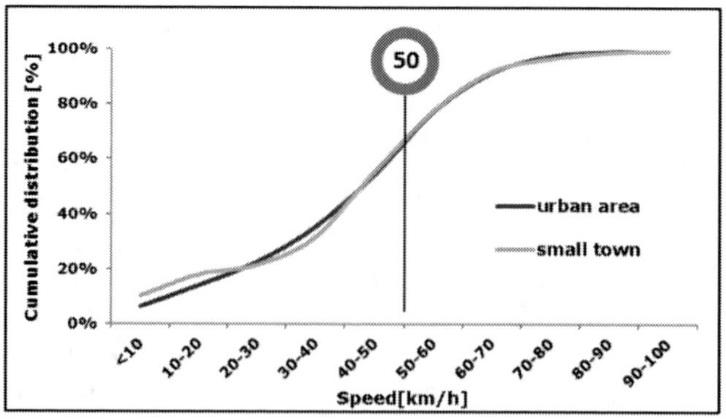

b)

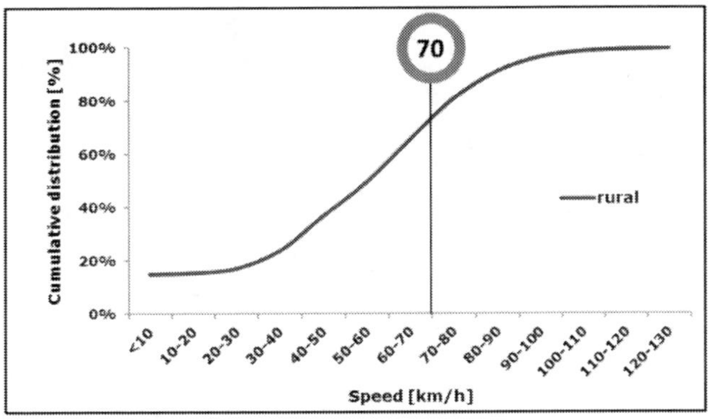

Fig. 3. Cumulative distribution of speed 10 m before the pedestrian crossing a) in towns and villages – V = 50km/h, b) non-urban areas – V=70 km/h [1].

a)

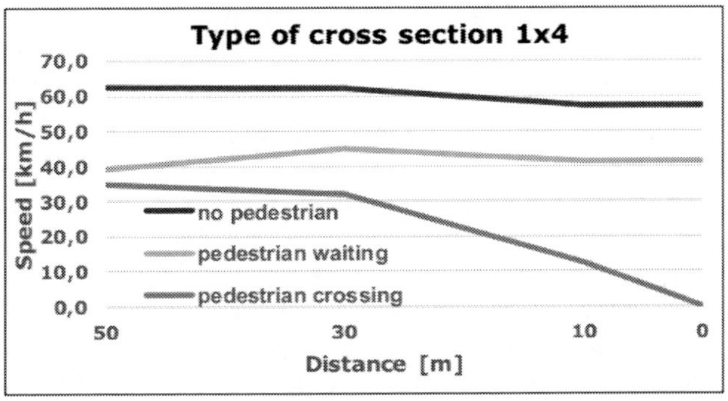

b)

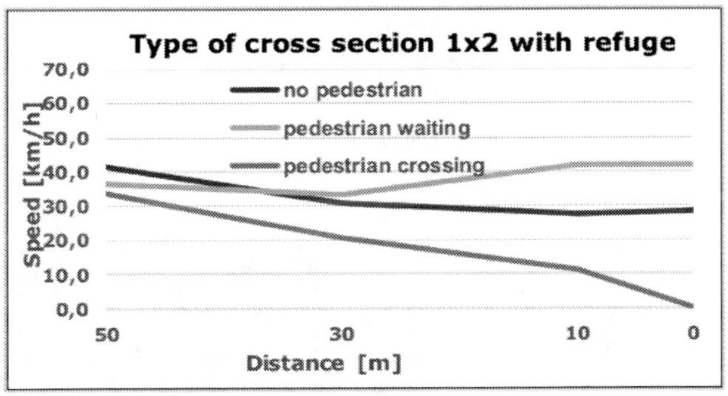

Fig. 4. Characteristics of vehicle speeds on approaching pedestrian crossings for selected cross-sections [1].

Seven and a half thousand drivers were surveyed. A clear majority of the respondents believed that you have to stop before the crossing if the pedestrian is before the kerb (89%) and when the pedestrian is already on the road (99.7%). They also believed that drivers fail to stop in these cases (60%). In selected test points drivers were observed to establish their real behaviour in situations pictured in the photos (Fig. 2). Survey results were compared against driver observations which showed that in reality many more drivers fail to stop and give way to a pedestrian who is either waiting to cross the road or has just moved onto the crossing (as opposed to what was declared in the survey) (Fig. 5).

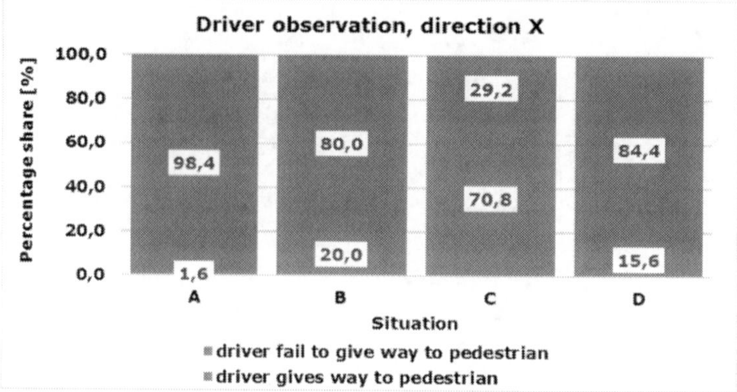

Fig. 5. Results of driver observations (Situations A, B, C, D - Fig.2) [1].

3 Selected recommendations for improving pedestrian safety

The analyses showed that pedestrians are most often put at risk by too long pedestrian crossings, vehicles going too fast around pedestrian crossings, lack of proper sight distance and poorly lit or unlit pedestrian crossings. The reason for such defective infrastructure is that planners, designers, contractors and maintenance services are not receiving any support from design, marking and maintenance regulations for pedestrian traffic. In addition, the Road Traffic Law is not restrictive enough when it comes to drivers' obligations towards pedestrian safety.

First, a method must be introduced to ensure a match between type of pedestrian crossing and speed limit. Because speed limits are generally exceeded everywhere in the country, it is recommended to use the percentile V_{85} of measured speed. A stock-taking of the infrastructure showed that many pedestrian crossings have 50, 60, 70 km/h speed limits and as much as 90 km/h in non-built-up areas. This applies to crossings that stretch over two lanes and more. The Pedestrian Safety Handbook [12] points out (Fig. 6) that a safe speed is 30 km/h. Speeds between 30 to 50 km/h are moderately safe. The relation between the probability of a vehicle hitting a pedestrian and the probability of pedestrian death shows a 50% chance of survival when the pedestrian is hit by a vehicle going at about 50 km/h. Thus, anything above 50 km/h is considered dangerous and speeds above 70 km/h are considered critical. This should be the basis for planning types of pedestrian crossings.

Fig. 6. Speed classification for pedestrian safety [2].

Polish design regulations [13] allow long pedestrian crossings up to four lanes in one direction or three lanes in two directions irrespective of traffic control and speed limits. Pedestrian crossings should be kept at a maximum of three lanes. There is nothing in the design regulations about the required driver-pedestrian sight distance. Neither does the Road Traffic Law [14] (art. 49.1.2) help engineers with that. It is legal to park vehicles within 10 m of a pedestrian crossing which does not guarantee the necessary sight distance. Drivers must be able to see a pedestrian waiting or stepping onto the crossing from a distance that will help them come to a stop safely. It is safer to follow the principle of providing adequate pedestrian sight distance. A pedestrian crossing the road should be able to judge thanks to the sight distance that there is a safe time gap to allow them to cross safely (Fig. 7).

a) pedestrian sight distance

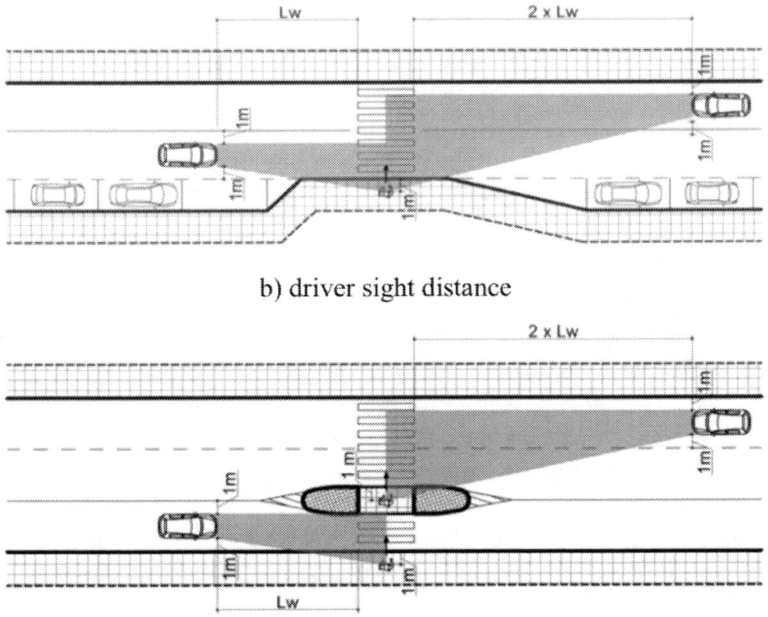

Fig. 7. Proposed principles for determining sight distance on pedestrian crossings [12].

Good sight distance L_w of a vehicle coming from the right or left side of a pedestrian is determined depending on: percentile V_{85} measured within the area of the pedestrian crossing, width of the pedestrian crossing (number of lanes to be covered by the pedestrian) and pedestrian speed V_p. Initial tests have shown that optimal vehicle sight distance for percentile V_{85}=50 km/h is 45 m.

To improve pedestrian safety on crossings, long crossings must be related to the volume of motorised traffic, pedestrian traffic and the actual speed of vehicles and sight distances. Safe speed cannot be higher than 50 km/h. In addition, clearance for parking must be checked in relation to the speed limit to ensure proper sight distance. Road safety management must be implemented [15,16] to provide pedestrian safety when designing and using road infrastructure.

4 Conclusions

Walking plays a crucial role in the transport system. This is true of small towns and villages with very little public transport and of big cities where walking is often used to move around the city. Walking is also part of many people's everyday lives (especially children, school youth, older people, people who do not have a car). Having said this, pedestrians are the most vulnerable road users and most at risk of death in road accidents, representing more than 30% of all road accident fatalities in Poland.

The relations between walking and driving and the hazards generated should be further researched to ensure that pedestrians can use roads safely.

The pedestrian-driver relation and the behaviour must be monitored on an on-going basis and pedestrian crossings must be inspected for safety. This will help improve pedestrian safety effectively.

References

1. K. Jamroz, M. Budzynski., S. Gaca, M. Kiec, et al., Methodologies for systematic studies of pedestrian behaviour and pedestrian-driver relations (Secretariat of the National Road Safety Council, 2015)
2. I.P. Alonso, D.F. Llorca, M.A. Sotelo, et al., Combination of Feature Extraction Methods for SVM Pedestrian Detection, IEEE. **8** (2007)
3. K. Diependale, Do pedestrians comply with traffic lights in Belgian cities?, in: ICTCT Work., (Ashdod, Israel, 2015)
4. B. Fildes, S. Lee, D. Kenny, W. Foddy, Survey of older road users: behavioural & travel issues - Report #57 (Monash University Accident Research Centre, 1994)
5. P.E. Gårder, The impact of speed and other variables on pedestrian safety in Maine, Accid. Anal. Prev. **36** pp. 533–542 (2004). doi:10.1016/S0001-4575(03)00059-9
6. J. Hatfield, R. Fernandes, R.F.S. Job, K. Smith, Misunderstanding of right-of-way rules at various pedestrian crossing types: Observational study and survey, Accid. Anal. Prev. **39** pp. 833–842 (2007). doi:10.1016/j.aap.2006.12.005
7. C. Johansson, Towards a method to improve road safety for pedestrians and cyclists especially in child pedestrian environments, Luleå University of Technology, Department of Civil, Environmental and Natural Resources Engineering, Architecture and Water, 2001
8. P.Olszewski, I.Buttler, W.Czajkowski, P.Dąbkowski, C.Kraśkiewicz, P.Szagała, A.Zielińska, Pedestrian safety assessment with video analysis, Transp. Res. Procedia. **14** pp. 2044–2053 (2016)
9. N. Guéguen, S. Meineri, C. Eyssartier, A pedestrian's stare and drivers' stopping behavior: A field experiment at the pedestrian crossing, Saf. Sci. **75** pp. 87–89 (2015). doi:10.1016/j.ssci.2015.01.018
10. E. Papadimitriou, A. Theofilatos, G. Yannis, Patterns of pedestrian attitudes, perceptions and behaviour in Europe, Saf. Sci. **53** pp. 114–122 (2013). doi:10.1016/j.ssci.2012.09.008
11. P. Tomczuk, Pedestrian safety - luminance measurements at a pedestrian crossing, Constr. Archit. **13** (2014)
12. K. Jamroz, S. Gaca, L. Michalski, M. Kieć, M. Budzyński, L. Gumińska, W. Kustra, T. Mackun, I. Oskarbska, J. Rychlewska, A. Ryś, J. Wachnicka, J. Wierzbicka, Ochrona Pieszych. Podręcznik dla organizatorów ruchu pieszego (Protection of Pedestrians. Guidelines for pedestrian traffic organizers) (National Road Safety Council, Gdansk, Warsaw, Cracow, 2014)
13. Regulation of the Minister of Transport and Maritime Economy of 2 March 1999 on technical conditions for public roads and their location, OJ 1999 No. 43 Pos. 430 (Poland, 1999)
14. Act of 20 June 1997 Road Traffic Act no. 98 pos. 3113 (Poland, 1997)
15. K. Jamroz, M. Budzyński, W. Kustra, L. Michalski, S. Gaca, Tools for road infrastructure safety management - Polish experiences, in: Transp. Res. Procedia, pp. 730–739 (2014). doi:10.1016/j.trpro.2014.10.052
16. K. Jamroz, M. Kiec, W. Kustra, Protection of Pedestrians as the Key Action for Implementing Poland's Vision Zero, paper 15, 0691, in: TRB 94th Annu. Meet., (Washington DC, 2015)

The effects of selected factors on pedestrian crossings in urban areas

Lucyna Gumińska[1,*]

[1]Gdansk University of Technology, Faculty of Civil and Environmental Engineering, Gdansk, Poland

Abstract. Pedestrian crossings are designed to help pedestrians cross a road. There are at-grade pedestrian crossings with or without traffic lights and grade separated crossings such as subways and footbridges. Pedestrian crossings may be located next to a junction or on road sections between junctions. Where at-grade crossings are involved, pedestrians and motorists interact, which may lead to dangerous situations and road traffic conflicts. These mutual interactions between infrastructure users determine how pedestrian crossings operate. They also affect the operation of junctions, if located next to them. The article presents the most important factors that affect the operation of pedestrian crossings. Because of the multiplicity of factors, only one of the parameters of traffic factors is described in detail.

1 Introduction

Pedestrian crossings are designed to help pedestrians cross a road [1]. There are at-grade pedestrian crossings with or without traffic lights and grade separated crossings such as subways and footbridges [2]. Pedestrian crossings may be located next to a junction or on road sections between junctions. Where at-grade crossings are involved, pedestrians and motorists interact, which may lead to dangerous situations and road traffic conflicts [3]. These mutual interactions between infrastructure users determine how pedestrian crossings operate. They also affect the operation of junctions, if located next to them.

The article presents the most important factors that affect the operation of pedestrian crossings. Because of the multiplicity of factors, only one of the parameters of traffic factors is described in detail.

"The research is conducted under the FLOW project. The project received funding from the European Union's Horizon 2020 research and innovation programme under grant agreement No 635998. This article reflects only the author's view and the Agency is not responsible for any use that may be made of the information it contains"

2 Factors influencing the operation of pedestrian crossings

Pedestrian crossings should ensure that people can move across them safely, comfortably, efficiently and economically whilst minimising the negative environmental impacts. These objectives are mutually dependent. Efficiency here is a measure of quality, i.e. capacity. Safety is determined by traffic layout, geometry and pedestrian and driver visibility. Traffic conditions translate into lost time.

* Corresponding author: lucgumin@pg.gda.pl

Pedestrian safety analyses that compare Polish domestic data and those from other European countries show that Poland features a very high rate of pedestrian fatalities, in excess of 30% of all road deaths in 2015. This is shown in Figure 1. While there is a significant difference between the 1990s, with pedestrian mortality in Poland per 1 m population at app. 70 people, and the post-2010 data at 30 people (Fig.2), the figure continues to be extremely high. It is clear that more investigation into the causes and factors of this is necessary.

It is important to know that Poland does not conduct research on the hazards that pedestrians are exposed to in urban and rural areas. If it did, efforts could be undertaken to understand the effectiveness and accuracy of treatments. As set out in the National Road Safety Programme for the Years 2013 – 2020, implemented by the National Road Safety Council, pedestrian safety is a strategic action under the pillar Safety of People.

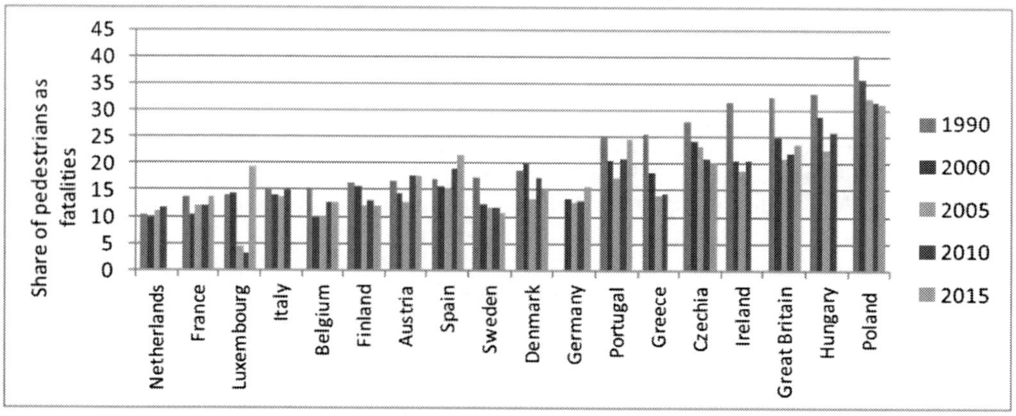

Fig. 1. Share of pedestrian fatalities, based on [4].

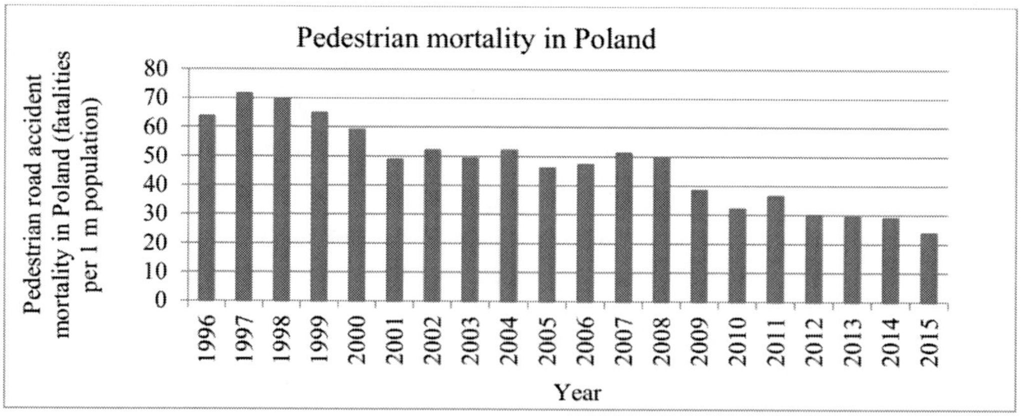

Fig. 2. Pedestrian accident mortality in Poland, based on [4,5].

Speed is a major factor in road accidents involving vehicles and pedestrians and determines accident severity. Fig. 3 shows the likelihood of a pedestrian death in relation to vehicle speed. Swedish researchers Erik Rsen and Ulrich Sander found that their predecessors based their findings on pedestrian serious injury or death which overestimated the risk of becoming a fatality [6,7]. Their research shows that the probability of pedestrian fatality is 100% for vehicle speed of app. 120 km/h, while other scientists such as R. Anderson claimed it is app. 57 km/h and E Pasanen, G. Davis and C. Oh at app. 100 km/h [8–10].

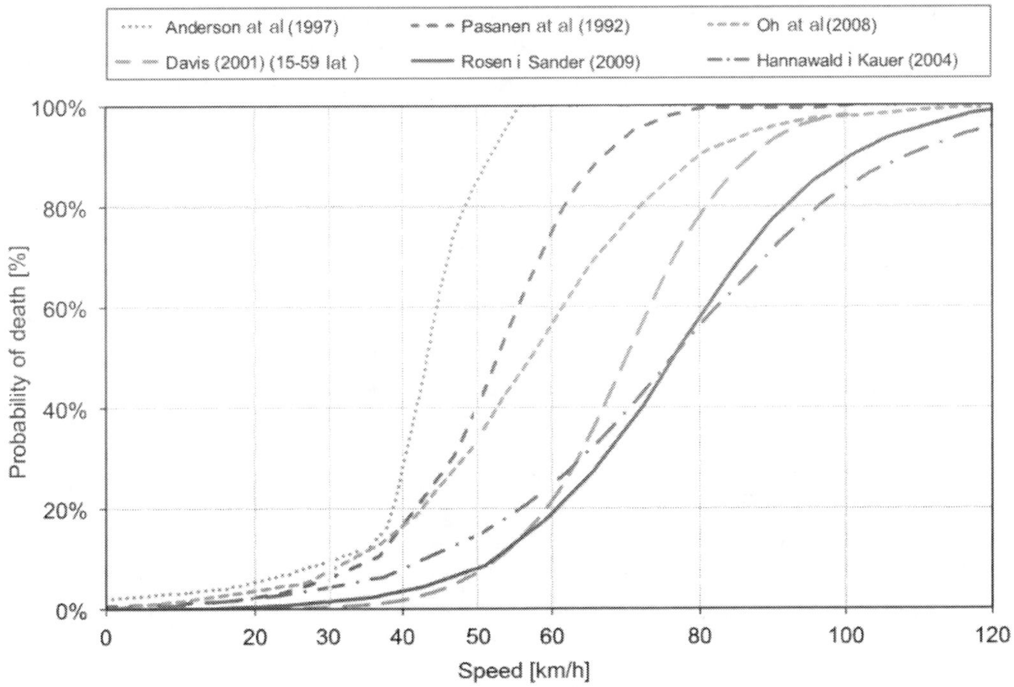

Fig. 3. Probability of pedestrian fatality, based on [7].

The hazards are quite extreme when there are conflicts involving pedestrians and vehicles. This happens when road user interactions are strong and may lead to an accident. If it is to be prevented, one of the road users must take up action to prevent the accident [11]:
- in the case of vehicles: brake, accelerate, change direction;
- in the case of pedestrians: stop, step back, jump to the side, walk faster, start running.

Traffic layout and traffic control around pedestrian crossings are also important [12]. Signalled pedestrian crossings usually allow vehicles to turn in the direction of the arrow on sign, when pedestrians have the green light on. Although they are required to stop before turning right, drivers do not always follow through which puts vulnerable road users at risk. By eliminating the possible driver-pedestrian collision, road safety will improve but the efficiency of the pedestrian crossing or junction is hampered.

Poor road user visibility is key to how a crossing or junction operate. Visibility is reduced as a result of road user behaviour and the dynamics of observer movement. It is important to keep space unobstructed on and around pedestrian crossings [1]. This issue is neglected in Poland, and the only available guidelines can be found in the Road Traffic Law [13] in Article 49.1.2 Vehicles must not stop "on pedestrian crossings, cyclist crossings and not less than 10 m before the crossing; where double carriageways are concerned, this ban also applies to areas after the crossing".

Key to our operational analyses are behavioural factors both regarding the driver and pedestrian. They comprise a set of mobility reactions which are a human body's response to specific incentives received by the road users through their senses [14]. A change in behaviour may be the result of learning from previously experienced incentives received by sight and hearing. Age and disability are also an important factor. All these features determine the speed of movement, response time, propensity for risky behaviour such as stepping onto the road or driving while the light is red or attempts to cross the street right before a vehicle.

Road elements are also important because they determine the behaviour, traffic parameters and pedestrian and driver responses. The location of the pedestrian crossing is equally important. If they are sited close to major traffic generators, it will affect pedestrian and driver volumes, traffic density, time lost and level of service.

There is a multiplicity of factors that determine the operation of pedestrian crossings. They include the weather, environment, lighting and legal regulations. Not all of them can be studied or analysed due to a lack of devices to measure the factors or missing or unreliable data sources. Considering the above, the most relevant factors are identified and divided into three groups:

1. behavioural factors – response time, age, disability;
2. traffic factors – traffic volume, speed of road users, traffic density, acceptable gaps, level of service (LOS), conflict situations;
3. road factors – pedestrian crossing geometry, road and pedestrian facility capacity, visibility of pedestrians and vehicles;

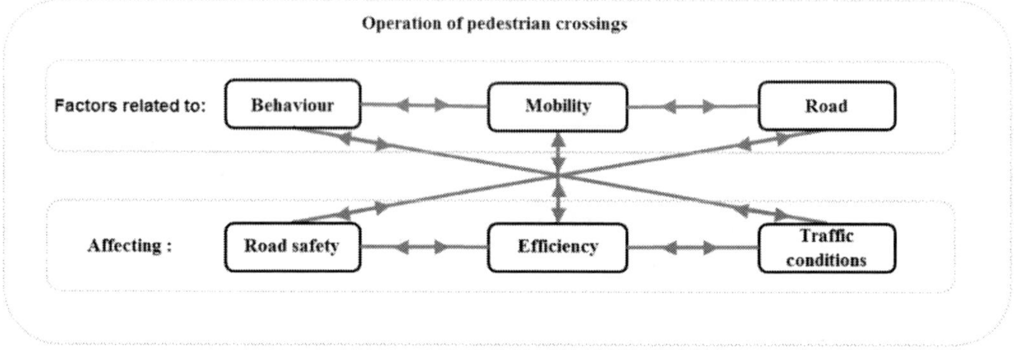

Fig. 4. Factors and features that affect the performance of pedestrian crossings.

3 Method for studying mean pedestrian walking speed

One of the parameters within the group of mobility factors is the mean walking speed of pedestrians on crossings. Specific sites were analysed to ensure data reliability. The study included:

- an inventory of pedestrian crossings under analysis to determine the average crossing distance;
- an identification of pedestrian crossing location including the generators of pedestrian traffic;
- video registration of pedestrian behaviour;
- measuring pedestrian volumes.

Four signalled pedestrian crossings were selected, each located on the main streets of Gdynia. Pedestrian behaviour was recorded using the Video Supervision System, part of the TRISTAR Traffic Control System [15,16]. It includes wide-angle cameras. The camera was placed at app. 4.5 m high giving a good view of the entire pedestrian crossing. Video registration was conducted during the afternoon peak (3.00 pm to 4.00 pm) in good weather. After the film material was collected, it was analysed by the observer.

4 Results

The video footage was evaluated. It helped to analyse the behaviour of some 870 pedestrians. The following characteristics were recorded: pedestrian volume, age divided into youth and adults (working age) and sex. Table 1 shows the results of walking speed analyses. The highest mean speed was recorded for men 5.2 km/h (1.45 m/s) and youth 5 km/h (1.40 m/s) and the lowest for older people 3.4 km/h (0.94 m/s). Mean speed for all groups under analysis was 4.9 km/h (1.36 m/s). The percentiles - 15th , 50th and 85th – were analysed and it was found that 15% of pedestrians in the analysed population walked at mean speed of up to 4 km/h (1.13 m/s), 50% of the pedestrians walked at mean speed of up to 4.8 km/h (1.35 m/s) and 15% at a speed higher than 5.6 km/h (1.57 m/s). Figure 5 shows the cumulative distribution of pedestrian speeds on the analysed pedestrian crossings.

Table 1. Mean pedestrian speed.

Age group	Sex	Number	Mean pedestrian speed		15th percentile [m/s]	50th percentile [m/s]	85th percentile [m/s]
			[m/s]	[km/h]			
Working age people	Women	412	1.34	4.83	1.14	1.34	1.55
	Men	324	1.45	5.22	1.26	1.41	1.66
Children	N/A	27	1.28	4.60	1.10	1.26	1.46
Youth	N/A	59	1.40	5.03	1.23	1.33	1.58
Older people	N/A	45	0.94	3.40	0.79	0.97	1.14
TOTAL		867	1.36	4.91	1.13	1.35	1.57

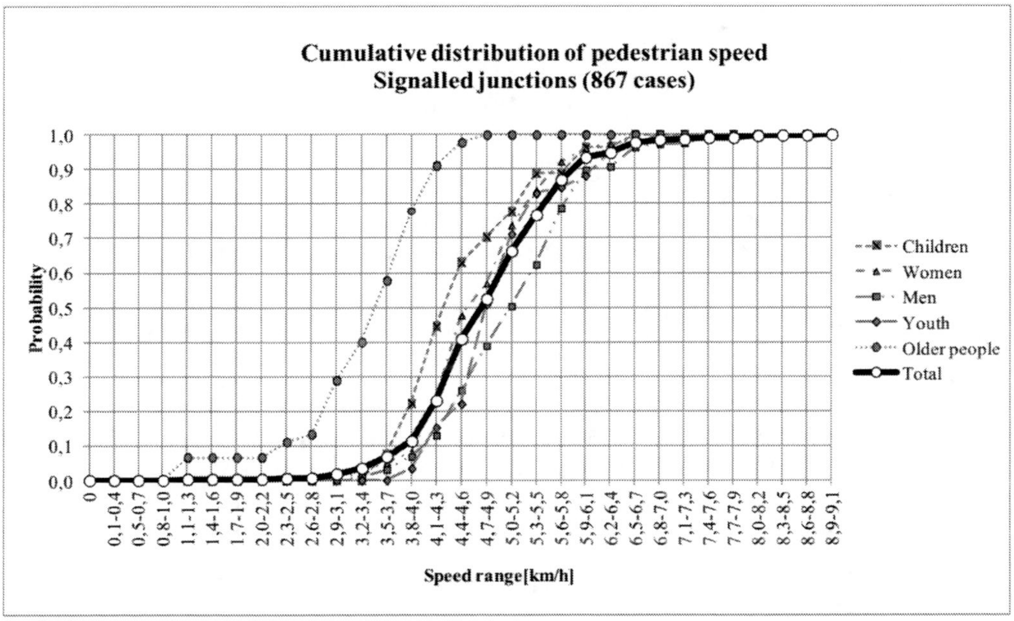

Fig. 4. Cumulative distribution of pedestrian speeds on signalled pedestrian crossings.

5 Discussion of the results in relation to the literature

Walking speed is one of the key parameters when designing and optimising the parameters of pedestrian crossing traffic lights. It depends on age, mobility motivation, destination and nature of trip, volume, disability, weather, type of terrain, location and geometry of the crossing, traffic layout, traffic control and, in the case of crossings with traffic lights, the part of green time duration which the pedestrian is using to cross the road.

In the 1980s US researchers [17] analysed the speed of pedestrians in free-flow using a sample of 967 pedestrians at a transport terminal in New York City. It was observed that 78% of the subjects moved at a speed less than 5 km/h (1.4 m/sec). Mean speed was 4.3 km/h (1.2 m/sec), with older men moving at 3.9 km/h (1.1 m/sec) and representing the 25th percentile of the cumulative distribution function. These results were compared to the study on Gdynia's signalled pedestrian crossings and presented in Fig. 5. As we can see, pedestrians from Gdynia sites moved slightly faster, which in the case of the 50th percentile was app. 0.2 km/h. The speeds of the pedestrian population studied can be described with normal distribution and standard deviation of 0.89.

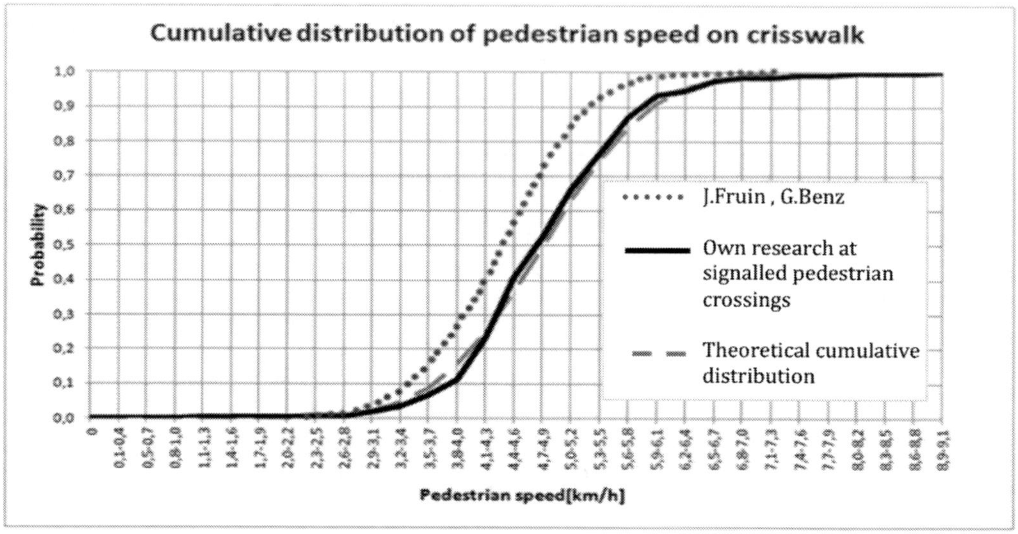

Fig. 5. Comparison of pedestrian speed distribution, based on [15].

6 Conclusions and recommendations for further research

More research should be done to include pedestrian crossings with different layouts and control systems. Further work will include the other parameters, for each group of factors that determine the performance of pedestrian crossings. This will later be used to develop mathematical models of pedestrian traffic. There are plans to conduct analyses using the micro-stimulation tool PTV Vissim with Viswalk module and compare the results with site test results. Finally, tools will be developed to help with planning, designing and auditing pedestrian facilities including a method for selecting the technical parameters and evaluating pedestrian crossings.

References

1. K. Jamroz, S. Gaca, L. Michalski, M. Kieć, M. Budzyński, L. Gumińska, W. Kustra, T. Mackun, I. Oskarbska, J. Rychlewska, A. Ryś, J. Wachnicka, J. Wierzbicka, Protection of Pedestrians. Guidelines for pedestrian traffic organizers (in Polish) (National Road Safety Council, Gdansk, Warsaw, Cracow, 2014)
2. Regulation of the Minister of Transport and Maritime Economy of 2 March 1999 on technical conditions for public roads and their location, OJ 1999 No. 43 Pos. 430 (Poland, 1999)

3. P.Olszewski, I.Buttler, W.Czajkowski, P.Dąbkowski, C.Kraśkiewicz, P.Szagała, A.Zielińska, Pedestrian safety assessment with video analysis, Transp. Res. Procedia. **14** pp. 2044–2053 (2016)
4. http://stats.oecd.org/#, (2016)
5. http://www.its.waw.pl/cms/domena_its/Piesi,0,3254,1.html, (2016)
6. E. Rosen, U. Sander, Pedestrian Fatality Risk as a Function of Car Impact Speed, Accid. Anal. Prev. **41** pp. 536–542 (2009)
7. J. Hu, K.D. Klinich, Toward designing pedestrian-friendly vehicles. Report no. Umtri-2012-19 (The University of Michigan, Transportation Research Institute, Michigan, 2012)
8. L. Hannawald, F. Kauer, ACEA Ehwal Effectiveness Study (Technische Universitat Dresden, 2004)
9. E. Roséna, H. Stigsonb, U. Sandera, Literature review of pedestrian fatality risk as a function of car impact speed (Sweden, 2011)
10. G. Davis, Relating severity of pedestrian injury to impact speed in vehicle-pedestrian crashes - Simple threshold model, TRB J. Transp. Res. Board. pp. 108–113 (2001)
11. R. Krystek, K. Jamroz, L. Michalski, R. Marszałek, Technika konfliktów ruchowych, (1991)
12. J. Oskarbski, L. Gumińska, M. Miszewski, I. Oskarbska, Analysis of signalized intersections in the context of pedestrian traffic, Transp. Res. Procedia. **14** pp. 2138–2147 (2016)
13. Act of 20 June 1997 - Road Traffic Act, OJ 1997 No. 98 Pos. 602 (Poland, 1997)
14. W.J. Broadus, *Behavior, An Introduction to Comparative Psychology* (Henry Holt and Company, New York, 1914)
15. J. Oskarbski., K. Jamroz, Reliability and safety as an objective of intelligent transport systems in urban areas, J. KONBiN. **2(34)** pp. 59–70 (2015). doi:10.1515/jok-2015-0024
16. J. Oskarbski, Perspectives of Telematics Implementation in Tri-City Transport Systems Management and Planning, Commun. Comput. Inf. Sci. **239** pp. 233–240 (2011)
17. C.V.Z. Brian, L. Bowman, J.J. Fruin, Planning Design and Maintenance of Pedestrian Facilities (Federal Highway Administration, 1989)

Risk assessment methodologies for pedestrian crossings without traffic lights – Warsaw case study – pedestrian safety assessment

Tomasz Mackun[1,*], *Artur* Ryś[1], and *Piotr* Tomczuk[2]

[1]Gdansk University of Technology, Faculty of Civil and Environmental Engineering, Gdansk, Poland
[2]Warsaw University of Technology, Faculty of Transport, Warsaw, Poland

Abstract. Based on data from 2015 [1], it was found that 31% of all fatalities in road accidents in Poland were pedestrians. In places accessible to pedestrian traffic 74.2% of total accidents involving pedestrians were recorded. Approximately 53.9% of accidents involving pedestrians take place in the proximity of pedestrian crossings. In the context of improving the safety of vulnerable road users, an assessment of the condition of infrastructure in this sensitive area is very important. Warsaw took up the challenge of a comprehensive assessment of pedestrian crossings in determining the level of road safety and lighting conditions. Research covered pedestrian crossings without traffic lights in three central districts of the city. The work included field research by teams of experts analysing the geometry of pedestrian crossings, their environment and user behaviour. To complete the task, methodologies for assessing risks to pedestrian safety were developed. In this article the authors have attempted to systematise a description of the method of pedestrian safety assessment in the area of pedestrian crossings.

1 Introduction

The development of road traffic in urban areas carries a number of risks to road users. Particular attention should be paid to road accidents and their consequences. In comparison with other European countries, Poland has a very high number of pedestrian fatalities [2]. Statistical analyses made available annually by the Police Headquarters confirmed these unfavourable data [1]. The main causes of accidents involving pedestrians indicated in the reports are: driver carelessness, pedestrian carelessness and lack of proper visibility of pedestrians. Despite the apparent decline in the number of accidents involving pedestrians [1] since 2000, they are still at risk in traffic. According to the analysis contained in the PH report [1] the highest number of accidents involving pedestrians occurs on the crossings. It would seem that the situation of pedestrians in Poland over the years has been gradually improving. However, a thorough analysis of a statistical report [1] shows an increase in the share of accidents at pedestrian crossings in relation to the total number of accidents per year in Poland.

According to the data [2], the number of pedestrian fatalities dropped by 9.7% (Fig.1) in the years 2010-2014 in Poland. The decrease in fatalities including pedestrians can be caused

* Corresponding author: tomasz.mackun@pg.gda.pl

by, among others, taking numerous actions to protect road users. A comparison of the rate of improvement in pedestrian safety in Poland and other countries puts Poland in a bad light. Many countries already by 2010 showed better indicators than Poland.

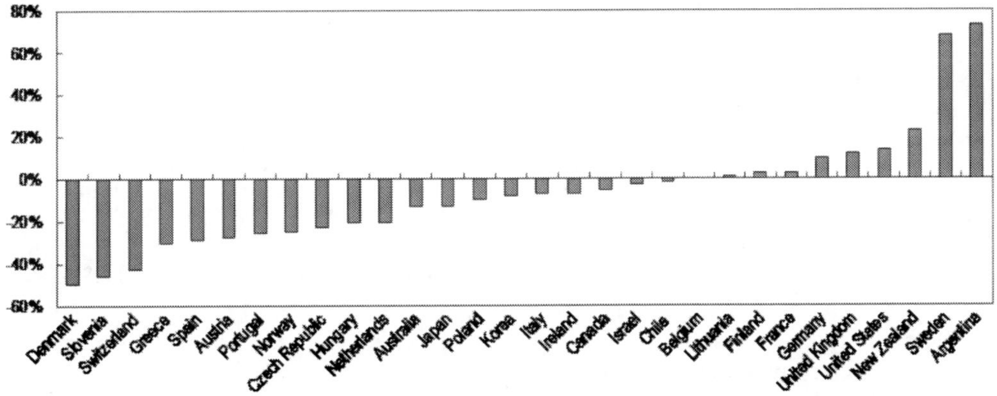

Fig. 1. Percentage change in the number of deaths among pedestrians, 2010-2014 [2].

The level of road safety depends on the users of the road infrastructure, its environment as well as on the technical condition of vehicles. A number of criteria that influence directly or indirectly the occurrence of accidents involving pedestrians in the vicinity of pedestrian crossings could be defined. Polish regulations do not specify how to determine visibility, procedures for selecting the type of pedestrian crossings are not adopted, maximum length of the crossing is not defined and there are no guidelines governing lighting conditions. Lack of these regulations means that pedestrian crossings are built that pose a high individual risk to pedestrians.

1.1 Overview

In the context of improving the safety of vulnerable road users in urban agglomerations, assessment of infrastructure is very important. The City of Warsaw commissioned an Audit of Road Safety and an assessment of lighting on 930 pedestrian crossings without traffic lights in three central districts of the city. The authors of this paper have attempted to systematise the process of assessment of the safety of existing crossings. It should be noted that the actions taken in the field of safety assessments were carried out consecutively with a team of lighting experts whose task was to assess the impact of lighting elements on the level of pedestrian safety.

Despite many efforts to improve the safety of pedestrians in Warsaw from 2011 to 2014 the number of accidents with pedestrians did not significantly decrease (434-490 cases/year). In 2015 there was a decrease to 390 accidents [3]. Because statistically they are a small number, pedestrian accidents are not mapped as "black spots". To act preventatively, a map of the level of risk at pedestrian crossings should be created regardless of whether accidents involving pedestrians took place there or not. Acting on the risk map the road manager can undertake preventative actions, rather than wait for accidents to happen.

It should be emphasised that so far Poland has not had similar research. A study of the literature [6][7] preceding conceptual and field work did not indicate a useful methodology. In the literature the authors encountered one source [6] where an attempt was made to create an objective numerical risk assessment of pedestrian safety, assigning weight factors and weights to individual parameters of a pedestrian crossing. However, they cannot be transferred from Italian to Polish conditions due to the specificity of the local behaviour of drivers and pedestrians and different traffic regulations. Therefore, a new, comprehensive assessment method had to be developed that would help to identify risk factors occurring at selected pedestrian crossings. The authors of the Polish assessment methodology decided to separate the assessment of the lighting parameters and other parameters of road safety.

2 Pedestrian risk assessment in the area of pedestrian crossings

2.1 Procedure

Step one – site inspection

The developed methodology of pedestrian risk assessment required the participation of experts in the field of road safety. In this particular case they were Road Safety Auditors certified by the Minister of Infrastructure and Construction after completing an appropriate course. The experience of the people assessing the risk is extremely important, because their assessment is based on knowledge and experience.

In the first step, an Auditor, Audit team or lead Auditor and their assistant inspect the site and complete a form with required data:

- accurate GPS location;
- the location of the crossing: on a straight section, on a curve, at a junction (on a priority road, on a minor road);
- surface: the type, condition;
- cross-section of road in the crossing area: the number of lanes, tram tracks, bike lanes, the existence of refuge islands;
- geometric parameters of crossing: length, width;
- the type of vehicles in the crossing area: trams, cars, bicycles;
- organization of traffic: traffic directions, number of lanes, the number of interconnections which pass through the pedestrian crossing;
- road signs related to pedestrian crossing markings: condition, type, completeness, speed limit;
- road markings: condition, crossing and pavement colour, crossing distinctness;
- water drainage: rain inlets in the crossing area, kerbside gutters, puddles;
- parking: distance from the crossing and the position of the vehicle parking (e.g. on the road, in a roadside lane, etc.),
- visibility: visibility measured in real conditions;
- subjective evaluation of the behaviour of users including pedestrians improperly crossing lanes, the speed of vehicles and the behaviour of drivers;
- photographic documentation.

Step two – risk identification

In the second step, on the basis of the collected data and observations the Auditor determines whether there are risks/irregularities. In the methodology, arising risks were predetermined due to:

- insufficient devices for people with disabilities and with reduced mobility:
    - lack of/inadequate equipment for the visually impaired,
    - lack of/insufficient ramp at the edge of the crossing.
- insufficient driver-pedestrian field of vision due to the reduction of visibility by:
    - parked vehicles,
    - fences, poles, advertising,
    - public transport stops,
    - greenery, other.

- irregularities in the road signage or markings due to incompleteness or poor technical condition,
- risks associated with the geometry of pedestrian crossing and road in the area due to:
    - excessive length of pedestrian crossing;
    - pedestrian crossing exceeding three lanes;
    - pedestrian crossing over excessively wide lanes;
    - use of "painted" surface which does not protect pedestrians;
    - insufficient width of the refuge island.
- risks arising from improper drainage in the pedestrian crossing area:
    - rain inlets in the area of pedestrian crossings;
    - lowest point of the vertical alignment/basin causing the formation of puddles.
- other risk factors:
    - bad technical condition of the roads, pedestrian crossings, pavements;
    - crossing prone to obstruction by vehicles;
    - unnecessary pedestrian crossing e.g. duplicating adjacent one;
    - lack of space for the accumulation of vehicles before the crossing.
    - other uncategorised above.

These risks are identified and recorded in a database. At this stage, the level of risk is not specified. The resulting database helps to sort and filter risk groups which can significantly facilitate the implementation of response programmes.

Step Three – defining the required field of view

In the third step of the procedure fields of view are defined that are required based on the speed limit and average speed values of the V85 of a particular cross-section. In order to determine the V85, data from the speed measurements were used and additional measurements were conducted.

Fig. 2. Risk identification due to limited field of view [8].

Step Four – formulation of recommendations

In the fourth step, the Auditor fills the database on the predefined recommendations aimed at improving the level of pedestrian safety and eradicating the risks. The resulting database helps to sort and filter groups of recommendations, which can significantly facilitate the introduction of response programmes.

Step Five – statement

In the fifth step, the Auditor prepares the statement describing the level of individual risk and indicating as many recommendations as possible aimed at improving the level of road safety.

Step Six - overall assessment

In the sixth step, the Auditor gives an overall assessment on a scale of 0 to 5, where the rating of 0 means the greatest risk for pedestrians and 5 lowest risk for pedestrians (Fig. 3).

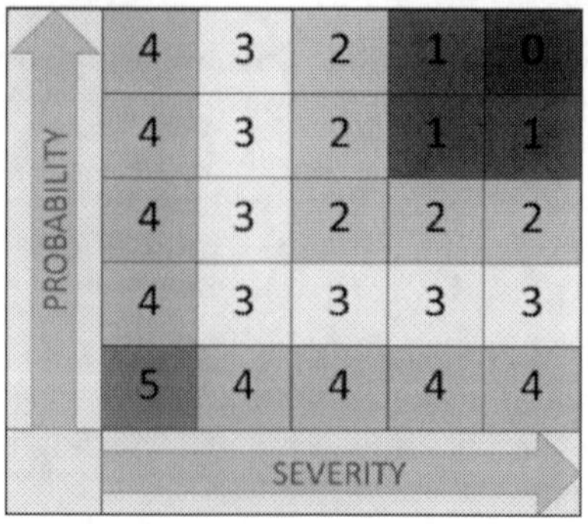

Fig. 3. Method of overall assessment of pedestrian risk in the area of pedestrian crossings.

Step Seven - verification

Due to the risk of subjectivity of each Auditor's evaluation, after the completion of the whole process of risk assessment, the material is evaluated by another party, i.e. the verifying Auditor. This person assesses the risks indicated, the adequacy of the observations and whether the recommendations are adequately and properly formulated. If the verifying Auditor disagrees with the assessment of a pedestrian crossing, a common position is found through discussion. If no agreement can be reached, a third Auditor is appointed who resolves the conflict and has a decisive vote. The procedure implemented in such a way should prevent subjective positions of an individual Auditor.

2.2 Additional data

Because not all data were available during the site inspection and seemed to be necessary for the assessment, efforts were made to obtain information about the road traffic volume (from a Warsaw transport model) and the traffic volume of pedestrians and cyclists by age group (own measurements conducted).

In addition, the method of required visibility was developed by the authors taking into account the position of a pedestrian at a distance of 1m from the edge of the road at drivers' speed of V85.

2.3 The resulting material

As a result of the work a collective database was created containing:
- a unique identification number given to each crossing;
- locations: streets, junctions, road class, number of lanes, description, district;
- assessment of the overall level of risk at a pedestrian crossing;
- a set of risks identified by type and cause;
- a set of recommendations by type;
- a hyperlink to the location of a crossing;
- traffic volume of vehicles;
- traffic volume of pedestrians and cyclists;
- a hyperlink to the individual worksheet of a pedestrian crossing.

In addition, for each crossing a worksheet is created, containing all the data from the site inspection, photographic documentation, assessment and statement by the Auditor.

2.4 Summary

The material enables the implementation of preventive measures. Pedestrian crossings can be filtered according to the ratings awarded. About 10% of the crossings with the worst assessment score were indicated where immediate corrective measures were deemed necessary. The database of risks and recommendations also helps to filter pedestrian crossings by specific problems such as visibility restrictions related to parking or inappropriate devices for the visually impaired.

The material, in the form of a database and specific worksheets, can be used to apply the method "from the general to the specific" which should be of significant benefit considering the high number of pedestrian crossings.

3 Conclusions

The authors took the decision not to link the assessment of lighting with the assessment of pedestrian safety due to the fact that in some cases the lighting expert's evaluation significantly differed from that of the road safety experts. If compiled in the form of a joint assessment, the data would statistically obscure some important problems.

Effective action to improve the safety of pedestrians should include, inter alia, conducting a systematic evaluation of safety at crossings and application of effective measures for the protection of pedestrians at the crossings where there is a particularly high pedestrian risk.

The authors aim to develop a coherent database of measurement results and expert assessments and information on road incidents involving pedestrians. Ultimately, it may be the base material for construction of a model quantifying the share of individual components in the overall level of pedestrian risk.

References

1. Road accidents in Poland in 2015. Report of General Police Headquarters of Poland (Warsaw, 2016)
2. Road Safety Annual Report 2016 (OECD Publishing, 2016). doi:10.1787/irtad-2016-en
3. Report on road safety in Warsaw in 2015 (Warsaw Municipal Road Authority ZDM, 2016)
4. P. Holec, T. Maixner, J. Skála, J. Smetana, J. Tesař, Posouzení přechodových svítidel, Společnost pro rozvoj veřejného osvětlení podzimní setkání, in: (Brno, 2009)
5. General Director for National Roads and Motorways, Manual for road safety auditors, part I and II (Warsaw, Poland, 2009)
6. A. Montella, F. Mauriello, Pedestrian crosswalks safety inspections: safety assessment procedure, in: 4th Int. Symp. Highw. Geom. Des., (2010)
7. D. Nabors, M. Gibbs, L. Sandt, S. Rocchi, E. Wilson, M. Lipinski, Pedestrian Road Safety Audit Guidelines and Prompt Lists, FHWA-SA-07-007 (2007)
8. K. Jamroz, S. Gaca, L. Michalski, M. Kieć, M. Budzyński, L. Gumińska, W. Kustra, T. Mackun, I. Oskarbska, J. Rychlewska, A. Ryś, J. Wachnicka, J. Wierzbicka, Protection of Pedestrians. Guidelines for pedestrian traffic organizers (in Polish) (National Road Safety Council, Gdansk, Warsaw, Cracow, 2014)

Pedestrian and bicycle bridges as examples of safe collision-free road crossings

Mikołaj Miśkiewicz[1*], *Łukasz* Pyrzowski[1], and *Romanika* Okraszewska[1]

[1]Gdansk University of Technology, Faculty of Civil and Environmental Engineering, Gdansk, Poland

Abstract. Pedestrians are most at risk when they are crossing the road. This represents a significant proportion of all fatalities among pedestrians, amounting respectively to 50% in non-built-up areas and 75% in built-up areas. The most frequent reason for this accident is failure to give way. What is most terrible is that 30% of pedestrian accidents occurred at marked pedestrian crossings. Therefore, an important part of pedestrian safety management is selecting the right type of crossings, which are suitable for the conditions. At certain speeds and traffic volume, the only safe option for pedestrian crossings is to apply multi-level solutions, that is footbridges or tunnels. The paper presents examples of infrastructure redesign by constructing footbridges and hence, separating pedestrians and vehicular traffic to improve pedestrian safety.

1 Introduction and regulations

A European Transport Safety Council report published in June 2016 [1] points out to Poland as a country in which the number of road deaths per travelled vehicle-distance is the highest in the European Union. In 2015, 26% of all car accidents involved pedestrians and 14% involved cyclists. Hitting a pedestrian is the most tragic kind of accident on Polish roads. 31.4% of fatalities are pedestrians. The largest number of pedestrian and cyclist accidents occurs in urban areas, however, the consequences of accidents which happen outside built-up areas are more tragic. More details about the situation on Polish roads may be found in the report of the Polish Traffic Police Service [2]. This statistic shows that with such poor safety of vulnerable road users, it is imperative that programmes and actions are developed to protect pedestrians and cyclists.

Pedestrians are most at risk when they are crossing the road. This represents a significant proportion of all fatalities among pedestrians, amounting respectively to 50% in non-built-up areas and 75% in built-up areas. The most frequent reason for this kind of accidents is failure to give way. What is most terrible is that 30% of pedestrian accidents occurred at marked pedestrian crossings (see Fig. 1) [3]. Therefore, an important part of pedestrian safety management is selecting the right type of crossings, which are suitable for the conditions. According to Polish law, pedestrians are allowed to cross the road as follows:

- in any place, if far from the designated pedestrian crossing (the distance to the nearest crossing should be longer than that required by road traffic law),

* Corresponding author: mmisk@pg.gda.pl

 DOI: 10.1051/matecconf/201712201005

- at-grade pedestrian crossings (with or without traffic lights; with an "island"; with barriers defining a path for pedestrians),
- collision-free grade-separated crossings in the form of a footbridge or tunnel.

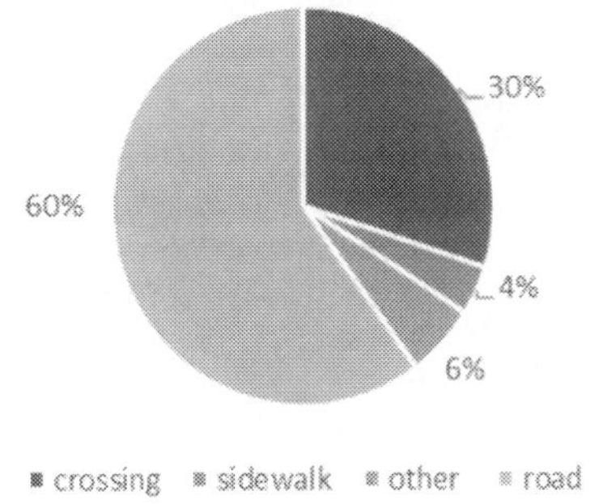

Fig. 1. Distribution of accidents with pedestrians on Poland's road network. Source: own study based on [3].

Pedestrians and cyclists prefer to use at-grade crossings. This is because the distance is shorter and there is no difference in elevation. However, in some cases, to ensure road user safety, grade-separated crossings are indispensable. Tunnels and footbridges should be used as collision-free crossings when:
- pedestrian routes intersect with higher class roads,
- pedestrian routes intersect with the roads of G (main road) or GP (major trunk road) class with heavy traffic,
- the location of an at-grade pedestrian crossing may pose a serious hazard to pedestrians or cause long delays for vehicles and pedestrians [3].

The right choice of pedestrian crossing must be preceded with an analysis of behavioural aspects of pedestrians. Studies show that, if the implementation of a grade-separated crossing extends the distance to be covered by pedestrians by 50% or more, pedestrians will try and cross the roadway despite the prohibitions [3]. Therefore, where possible, at-grade crossings should remain. However, as we can see from the nomogram in Fig. 2, for certain speeds and traffic volumes, grade-separated solutions such as footbridges or tunnels are the only safe option. This solution should be additionally supported with physical barriers to stop pedestrians from using prohibited at-grade crossings.

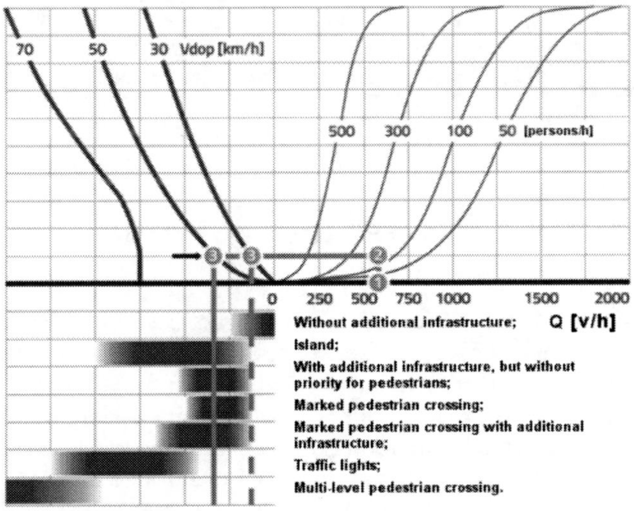

Fig. 2. Selection of the type of pedestrian crossings in urban areas. Source [3].

2 Design conditions for footbridges

The most popular static schemes applied in footbridges are beam, arch and cable-stayed structures. They should be characterized by a low value of the structural height parameter, due to the fact that many of them are built above the roads. The span length of footbridges is determined by obstacles. In the case of road crossings, the main parameter is road clearance. It depends on the road class given in national alignments[4]. For example, in Polish regulations for GP[i] class roads the minimum value of a traffic lane is 3.5 m, shoulder 0.5 m and verge 1.5 m. The usable width of footbridge decks is also regulated. Bidirectional decks for pedestrians should have the minimum width of 1.5 m and 2.0 m for cyclists. The structures in Poland must comply with the regulations given in [4] and [5].

3 Case studies

3.1 Footbridge in Męcikał

The first example of a footbridge which saves human lives is the footbridge in Męcikał over the Brda river (Fig. 3). The structure was built in 2010, the details may be found in [6]. The footbridge was built in response to the needs of local residents. It connects two places which are periodically visited by a significant number of people – a school and church. Before the new route was opened, people and cyclists were forced to use the narrow walkways on the bridge, along the road No. 235. The new footbridge is not only for the convenience of pedestrians but for their safety as well, because as we can see in Fig.1 4% of all accidents with pedestrians occur on sidewalks. The steel arch structure with a span of 42 m and full length of 58.46 m, improved safety of local pedestrians and cyclists, especially children walking to school.

Fig. 3. Footbridge in Męcikał: a) location, b) view from the road No. 235. Source: www.google.pl/maps.

3.2 Footbridge in Rzeszów along Piłsudskiego avenue

The footbridge in Rzeszów was built in 2012 over a junction of Z class road - Piłsudskiego avenue and L class road Grunwaldzka street (Fig. 4). The structure was erected to solve traffic problems in one of the busiest places in the city centre. This is an example where an at-grade pedestrian crossing is not a good solution due to heavy traffic. Crossing over the road was chosen over crossing under the road because of the extensive underground infrastructure and the enormous cost of the tunnel. The footbridge separates pedestrians and vehicular traffic. To ensure pedestrian access to the footbridge from every side and free movement in any direction, the round shape was proposed. The ring has a diameter of about 40 m and is divided into four spans, for details see [7]. The unusual shape of the footbridge makes it one of the few of this type in the world.

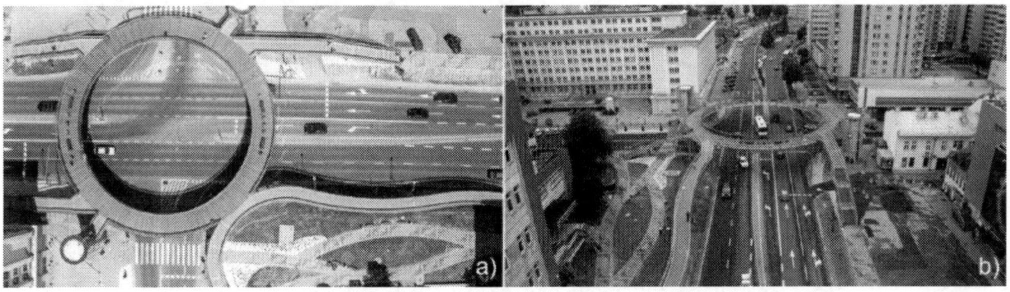

Fig. 4. Footbridge in Rzeszów on Piłsudskiego avenue. Source: www.skanska.pl.

3.3 Overpass for cyclists in Gdynia

The overpass for cyclists in Gdynia was opened in 2013 (Fig. 5). It is a multi-span steel structure of 301 m total length. It connects two streets Kontenerowa and Unruga. Cyclists may use this overpass to move along Kwiatkowskiego street where cycling is forbidden due to heavy traffic with a speed limit of 70 km/h. This calls for a segregation of cycle and motor traffic due to traffic volume.

Fig. 5. Overpass for cyclists in Gdynia.

3.4 Footbridge in Warsaw over S8 express road

An example of a pedestrian crossing over a higher class road (expressway) is the footbridge in Warsaw (Fig. 6). This is an example of an engineering structure where the design plays an important aesthetic role rather than mechanical principles. This structure is composed in acoustic screen tunnels and allows grade-separated pedestrian traffic between two districts of Ruda and Potok. The construction was finished in 2015. The footbridge is a steel double span arch with a composite concrete-steel deck. Its main dimensions are 21.95 + 17.85 m span length in the support axis and 6.36 m width [8]

Fig. 6. Footbridge in Warsaw over S8 express road during load test.

3.5 Suspension footbridge in Radom over Szarych Szeregów street

The footbridge in Radom links two sides of the Obozowisko district divided by a dual carriageway of Szarych Szeregów and provides access to a catholic parish church (Fig. 7). It replaces the traditional pedestrian crossing on a heavy traffic road. The structure's basic dimensions are: span of 40.2 m and 4.5 m width. The footbridge is a suspended structure with an interesting architectural form. The steel-concrete composite deck is suspended on one rope and hangers to two concrete towers.

Fig. 7. Suspension footbridge in Radom over Szarych Szeregów street.

3.6 GFRP composite footbridge

The last example can make a real revolution, when it comes to the availability of cheap and fast to assemble structures, which in a very easy way may significantly increase the safety of vulnerable road users. The fully polymer composite footbridge presented in Fig. 8 was constructed by the Fobridge consortium in 2015, as a research object [9][10][11]. Numerous analyses and tests confirmed that this novel structure can be applied over roads. This single span footbridge length can be up to 22 m long. The span was tested at the Gdansk University of Technology at 14 m. It is enough to cross a dual carriageway. The structure is durable, dynamically resistant, incombustible, easy to install and maintain, resistant to weather conditions and also aesthetically interesting. To produce the footbridge, the environmentally friendly PET foam core can be used. It may be sourced from recycled plastic packages and produced with lower energy consumption and much less CO_2 emissions. It is a very attractive proposal for applications over roads and obstacles. Additionally, its very short production time and lightness, speed up the construction. As you can see in Fig. 9, the proposed footbridge ensures that a road with up to three lanes can be crossed safely.

Fig. 8. GFRP footbridge constructed within Fobridge project.

Fig. 9. Visualization of Fobridge GFRP footbridge over a three lane road.

4 Conclusions

The principles and criteria for the choice of pedestrian crossing types are described in [3]. The author claims that to reduce the probability of pedestrian accidents the sources of danger related to time and space should be eliminated. This is why traffic lights, asylum islands, footbridges and tunnels must be constructed to move vulnerable road users to a different level than the road. Where justified, pedestrian bridges may be the best or only option available.

The article presents examples of grade separated pedestrian crossings using different types of footbridges. The examples show the variety of forms and the space-making potential of engineering structures [12]. The main advantage of pedestrian footbridges is that they separate pedestrians from road traffic. As a result, footbridges (and stopping pedestrians from crossing the roadway at-grade) may reduce pedestrian accidents up to 90%. The main drawback of collision-free crossings are their costs and their invasive character. The last example of a GFRP structure makes a faster, cheaper and environmentally friendly redesign of the infrastructure possible [9][11]. Lightweight GFRP structures do not need extensive foundations, they are price competitive, do not need expensive maintenance and their manufacture, transportation and installation are much easier and faster compared to steel or concrete solutions.

References

1. Ranking EU Progress on Road Safety, 10th Road Safety Performance Index Report, available at: http://etsc.eu/wp-content/uploads/10-PIN-annual-report-FINAL.pdf (2016)
2. E. Symon, Road accidents in Poland in 2015. Report of General Police Headquarters of Poland (Warsaw, 2016)
3. K. Jamroz, S. Gaca, L. Michalski, M. Kieć, M. Budzyński, L. Gumińska, W. Kustra, T. Mackun, I. Oskarbska, J. Rychlewska, A. Ryś, J. Wachnicka, J. Wierzbicka, Protection

of Pedestrians. Guidelines for pedestrian traffic organizers (in Polish) (National Road Safety Council, Gdansk, Warsaw, Cracow, 2014)

4. Regulation of the Minister of Transport and Maritime Economy of 2 March 1999 on technical conditions for public roads and their location, OJ 1999 No. 43 Pos. 430 (Poland, 1999)
5. Regulation of the Minister of Transport and Maritime Economy of 30 May 2000 on technical conditions, which should be ful-filled in case of the road engineering objects and their placement, OJ 2000 No. 63, Pos. 735 (Poland, 2000)
6. M. Malinowski, M. Miśkiewicz, J. Chróścielewski, A. Banaś, Arch footbridge under load testing, Zesz. Nauk. Politech. Rzesz. Bud. I Inżynieria Środowiska. **58** pp. 165–172 (2011)
7. T. Siwowski, A. Wysocki, The circular footbridge in a city center, in: 4th Int. Conf. FOOTBRIDGE, pp. 713–718 (Wroclaw, Poland, 2011)
8. Ł. Pyrzowski, J. Kałuża, M. Miśkiewicz, Ł. Filar, M. Wazowski, K. Wilde, Load tests of arch footbridge 10t over road S8 in Warsaw, in: Wrocławskie Dni Most. Most. Łukowe - Dzieła Kult. Proj. Budowa, Utrzymanie, pp. 419–426 (Wroclaw, Poland, 2015)
9. R. Okraszewska, L. Pyrzowski, M. Miskiewicz, Composite footbridge - synergy effect in cooperation between universities and industry, in: ICERI2014 Proc., pp. 2897–2903 (2014)
10. M. Miskiewicz, L. Pyrzowski, J. Chroscielewski, K. Wilde, Structural health monitoring of composite shell footbridge for its design validation, in: 2016 Balt. Geod. Congr. (BGC Geomatics), pp. 228–233 (IEEE, 2016). doi:10.1109/BGC.Geomatics.2016.48
11. J. Chróścielewski, M. Miśkiewicz, Ł. Pyrzowski, K. Wilde, Composite GFRP U-Shaped Footbridge, Polish Marit. Res. **24** pp. 25–31 (2017). doi:10.1515/pomr-2017-0017
12. M. Salamak, K. Fross, Bridges in Urban Planning and Architectural Culture, Procedia Eng. pp. 207–212 (2016)

MATEC Web of Conferences **122**, 01006 (2017) DOI: 10.1051/matecconf/201712201006
GAMBIT 2016

Growing role of walking and cycling and the associated risks

Romanika Okraszewska[1,*], *Krystian* Birr[1], *Lucyna* Gumińska[1], and *Lech* Michalski[1]

[1]Gdansk University of Technology, Faculty of Civil and Environmental Engineering, Gdansk, Poland

Abstract. Increasing the role of active mobility, including walking and cycling, is one of the tools for developing sustainable urban transport systems as recommended by the EU. The article describes the trends in pedestrian and bicycle traffic in Poland and its share in urban modal split. It identifies and describes the main sources of risks to pedestrian and cycle safety. Recommendations are proposed on how to ensure that pedestrians and cyclists can use traffic safely.

1 Introduction

Tackling climate change is one of the biggest challenges for international cooperation [1]. It is largely influenced by transport which represents a major and increasingly big source of greenhouse gas emissions [2]. Urban traffic accounts for 40% of CO_2 emissions and 70% of other road transport-related emissions [3]. In a bid to tackle climate change the European Commission has set emission targets. They are to reduce greenhouse gas emissions by 2050 by at least 60% compared to 1990 levels [4]. As regards transport, the European Commission believes that the targets can be achieved through progress in technology and/or change in how the public use transport, i.e. by reducing car trips in favour of public transport and choosing active forms of mobility such as walking and cycling [5]. If we could increase the share of walking and cycling in the modal split by reducing car trips, we could also reduce congestion which affects all of Poland's major cities and people's quality of life. Active forms of mobility are also good for those walking or cycling and they have an indirect positive effect on the health of everyone else [6]. Apart from those health benefits, non-motorised forms of transport also offer environmental, economic, social and spatial benefits. This is why many cities in the world and in Poland have policies designed to increase the share of active forms of mobility in how people travel [5]. Because urban and rural walking and cycling differ, this article covers built-up areas only. It gives an overview of the trends and the changing role of pedestrian and cycle traffic in Poland's built-up areas between 2009 and 2016. As well as identifying the risks to pedestrians and cyclists, the article uses international experience and available cyclist and pedestrian accident data to discuss the relation between the safety of vulnerable road users and the volume of pedestrian and cycle traffic.

* Corresponding author: rokrasze@pg.gda.pl

2 Trends in pedestrian and cycle traffic

Walking is the oldest form of transport. Today, walking is usually used for short trips. Up to 60% of trips up to 2 kilometres are made on foot and as many as 80% for trips up to 1 kilometre. Studies of transport behaviour conducted in Poland show that walking accounts for 10-29% (Table 1) of all urban trips. In some European cities such as Lyon (France), Oslo (Norway), Helsinki (Finland) and Valencia (Spain) walking has a share of more than 30% [6]. The studies also established that cycle trips differ from city to city and range between 1% to 10% of all urban trips (Table).

Table 1. Modal split in selected Polish cities. Source: [6].

City	Year	Walking	Cycling	Private transport	Public transport
Chorzów	2011	11%	10%	56%	23%
Gdańsk	2016	21%	6%	41%	32%
Poznań	2013	13%	4%	39%	43%
Wrocław	2011	19%	4%	42%	35%
Płock	2015	10%	2%	53%	35%
Kraków	2013	29%	1%	27%	43%
Szczecin	2010	19%	1%	43%	35%
Warsaw	2010	21%	1%	24%	54%
Kraków	2010	25%	1%	28%	46%

As we can see in Table 2 cycling has been growing annually whilst walking maintains a fairly constant share in the modal split in Polish cities. Analysis of the change in the modal split suggests that a large part of cyclists give up their daily trips by public transport. In the last seven years Gdansk's cycle traffic has grown by 4 percentage points whilst the share of public transport dropped by 6 percentage points [7,8].

Table 2. Change in cycling and walking within the modal split of selected Polish cities. Source: [7–12].

City	Trips		Trips	
	Walking	Cycling	Walking	Cycling
Cracow	2003		2013	
	29%	1%	30%	2%
Warsaw	2005		2016	
	21%	1%	18%	3%
Gdansk	2009		2016	
	21%	2%	21%	6%

Depending on the methodology, transport behaviour studies define travelling on foot as trips not shorter than e.g. 250 m [9,10,13]. When treated as an element of public transport trips, walking and cycling are interpreted differently depending on the methodology.
Designed to increase active mobility and reduce motor traffic, many strategies for sustainable transport foster cycling and walking by curbing car space. This includes limiting parking

space, introducing traffic restrictions and traffic calming and banning cars entirely from specific streets. Another objective is to develop pedestrian and cyclist infrastructure. Warsaw's total length of cycle routes is 457 km, 219 km in Gdansk, 165 km in Krakow, 164 km in Łódź, 107 km in Białystok and 30 km in Słupsk. If we consider the relation between the length of cycle routes [km] and the size of the city [km^2], Białystok comes up first with a rate of 1.05, followed by Warsaw at 0.88, Gdansk at 0.84, Słupsk at 0.70, Krakow at 0.50 and Łódź at 0.49.

Due to the scale of change in cycle traffic, the growth can be seen not only in the statistics but on the streets as well. The growing significance of cycling as part of the modal split is accompanied by an increase in cyclists. As an example, Gdansk monitors cycle traffic on specific cross-sections using automatic counters [14] (Fig. 1). There are also regular manual counts conducted by the staff of the Gdansk Development Agency [15].

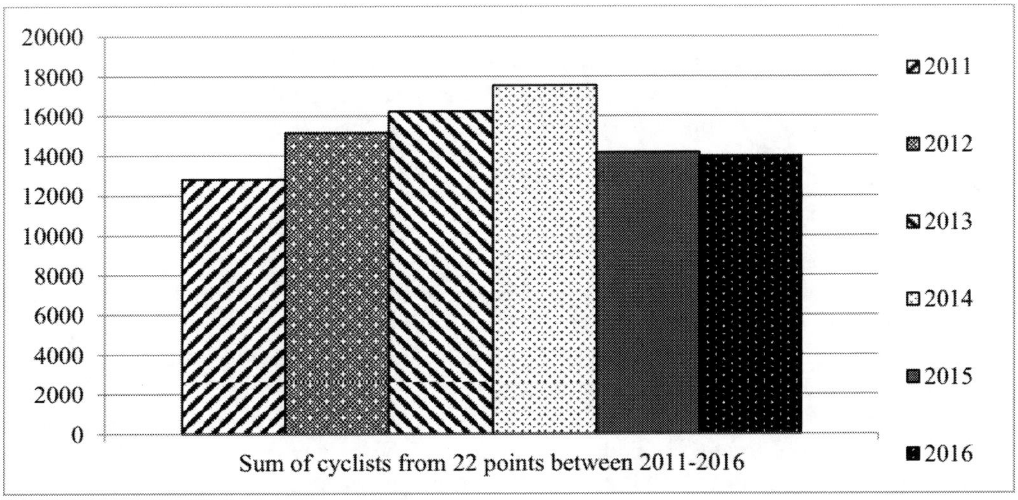

Fig. 1. Sum of cyclists recorded between 2011-2016 by automatic traffic counters on 22 cross-sections. Source: own work based on [14].

The results of automatic and manual counts are consistent and show an upward trend in cyclists. In 2015 the number of cyclists dropped on the analysed cross-sections while cycling gained a higher share in the modal split. This could be a consequence of new cycling infrastructure projects leading to a more dispersed cycle traffic on the growing network of cycle routes.

3 Road safety level

The change in pedestrian and cycle traffic affects the safety of vulnerable road users. Overall, Poland's road safety has improved significantly over the last five years (Fig.3). In 2015 compared to 2010 road accidents dropped by 15% (5 865 accidents), casualties went down by 19% (9 174 people) with fatalities decreasing by as much as 25% (970 people). Pedestrian safety has improved as well. Pedestrian accidents in the last five years dropped by nearly 24% with fatalities down by 26%. Cycle accidents, on the other hand, are on the rise. In 2015 compared to 2010 cyclist accidents went up by 18.3%, casualties by nearly 17% and fatalities increased by 7%.

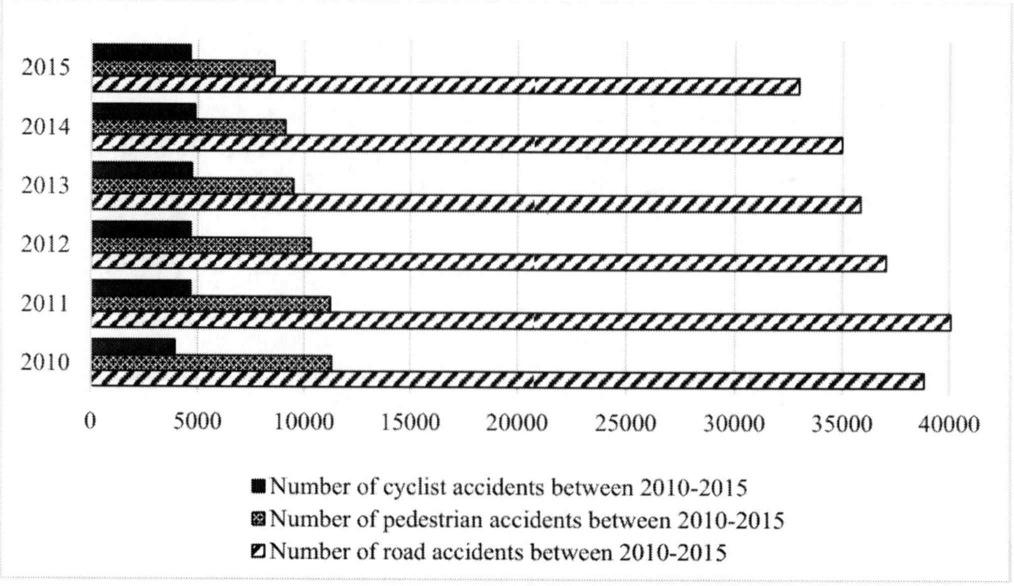

Fig. 2. Number of accidents by type of casualties. Source: own work based on National Police annual reports 2010-2015.

The risks to pedestrian traffic can be divided into groups involving:

- dangerous driver behaviour that may lead to a pedestrian accident (i.e. failure to give way, driving too fast, running the red light),
- dangerous pedestrian behaviour (i.e. stepping onto the road suddenly, using the wrong side of the road if there is no pedestrian infrastructure),
- poor pedestrian infrastructure (i.e. no pedestrian pavements, narrow pavements, obstacles on pavements, poor lighting of pedestrian facilities, conflicts with cyclists),
- faulty pedestrian facilities (i.e. pedestrian crossings located in front of bus bays, no pavements).

Pedestrian risks are covered in detail in the manual for pedestrian traffic designers titled "Pedestrian Safety" [16]. The book presents methods for assessing pedestrian risks and proposed pedestrian safety measures.

Safety is a key factor when deciding to use the cycle as a means of transport. Unless cyclists feel safe, they are less likely to use the cycle as a means of transport. Police reports provide cycle accident statistics. They can be used to understand the conditions and circumstances that determine cycle safety. The factors identified in the reports can be grouped into main categories which involve:

- external conditions (time and circumstances of the accident, time of day, lighting, cyclists using lights and reflectors, the weather),
- the behaviour of road users,
- infrastructure and layout.

The majority of cycle accidents:

- occur in daylight and do not involve alcohol,
- involve side crashes, followed by rear and head-on collisions,
- involve cars, followed by lorries;
- occur when cyclists or drivers fail to give way, when drivers overtake illegally and cyclists turn illegally.

Police statistics shows that drivers and cyclists take equal blame for accidents. The most frequent cause is wrong behaviour by one of the road users. Drivers will usually fail to give way, overtake and turn illegally, go too fast for the conditions, bypass illegally and do not keep the right clearance. Cyclists, on the other hand, cause accidents by riding on the wrong side of the road and riding across pedestrian crossings illegally [16]. The age distribution of cyclists involved in accidents shows that accidents most often involve young cyclists aged 11-19 and those aged 50 to 59.

4 The level of risk to pedestrians and cyclists in Poland

The level of cycle and pedestrian traffic safety is determined depending on:
- the area (e.g. number of fatalities per one million kilometres),
- demographics (e.g. share of pedestrians/cyclists in total road accident fatalities or the number of pedestrian and cyclist fatalities per one million population),
- transport (number of fatalities in cyclist accidents per 100 km of cycling).

For a number of years Poland has been the European Union's worst country for pedestrian safety Fig. 3 and cyclist safety Fig. 4 in road traffic.

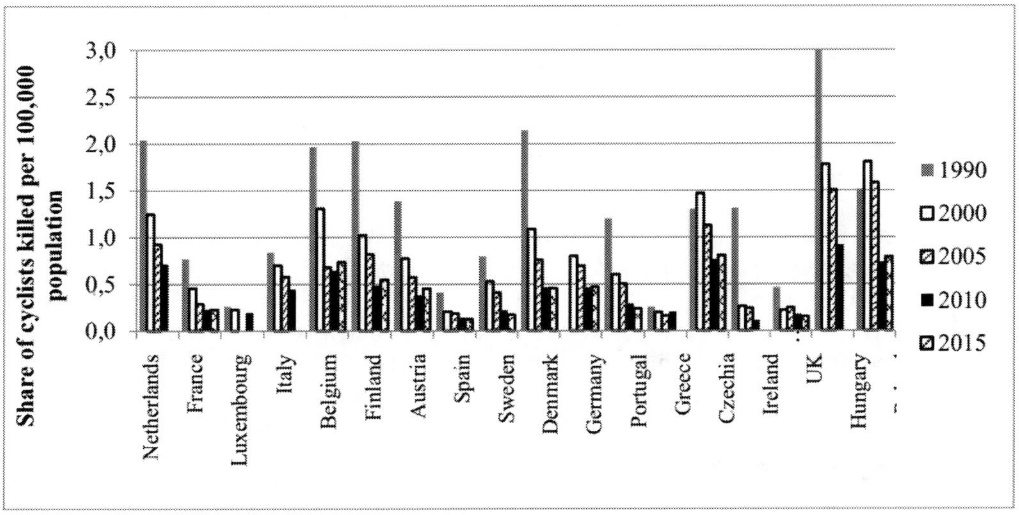

Fig. 3. Share of cyclists killed per 100 000 population. Source: own work based on [17].

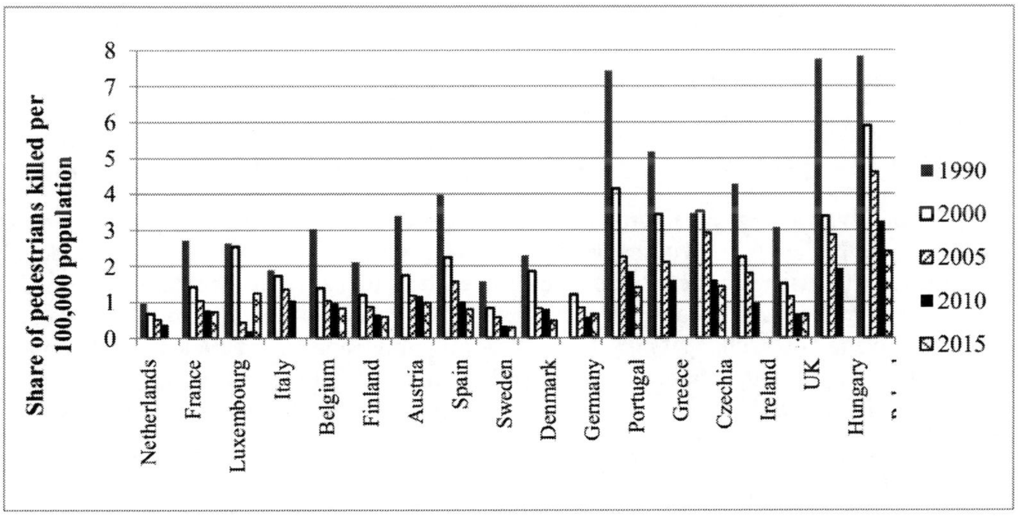

Fig. 4. Share of pedestrians killed per 100 000 population. Source: own work based on [17].

The country's pedestrian and cyclist safety rates vary across the country and depend on the administrative area's characteristics. They also depend on the degree of urbanisation. Built-up and non-built-up areas have different pedestrian and cyclist safety rates [19]. While the majority of pedestrian and cyclist accidents occur in built-up areas, the consequences of rural accidents are much more severe. In 2015 there was one pedestrian fatality in every third accident in a non-built-up area and one cyclist fatality in every fifth compared to one pedestrian fatality in every thirteenth accident in a built-up area and one cyclist fatality in every twentieth accident [20].

However, none of these rates give a realistic measure of cycling safety or directly show the risk of an accident when cycling or walking. The number of pedestrian and cyclist accidents which is the basis for calculating all rates, depends on a number of factors [16,17] including traffic volume. Using statistics from US cities P.L. Jacobsen expressed the relation between the number of accidents and how big cycle traffic is using this formula [21]:

$$\textit{Number of cyclist accidents} = \textit{cycle traffic volume}^{\,0.4} \tag{1}$$

The share of cycling in Poland is still low. The authors of the Strategy for Transport and Mobility in Gdansk's Metropolitan Area studied the relation between the number of cyclist accidents and the low share of cycling in overall trips. The results show that cycling carries the highest risk [22].

5 Conclusions

The role of walking and cycling as part of urban transport services is growing. This is the result of efforts to promote cycling, foster public transport and limit car use.

Over the last five years Poland's pedestrian safety has improved. The number of pedestrian accidents dropped by nearly 24% and the number of killed went down by 26%. Despite that in Poland every third road fatality is a pedestrian compared to the EU's every fifth. This is the result of driver behaviour and condition of pedestrian facilities.

Over the same period **the number of cyclist accidents went up by 18.3 %,** the number of casualties rising by nearly 17% and fatalities by 7%. Forecasts show that cyclist accidents will continue to increase. This is due to growing cycling traffic, driver and cyclist behaviour, the safety standards of cycle infrastructure and how cycle traffic is organised.

The road safety system today still fails to provide an effective way to protect pedestrians and cyclists from injury or death. While cities are building cycle infrastructure, it is not available outside them. The new infrastructure does not come with **campaigns to promote safe behaviour** mainly that of cyclists using road traffic.

6 Systemic solutions to improve pedestrian and cyclist safety

To address the challenges of growing cycle trips and the potential increase in cyclist accidents, actions are needed to reduce the impacts as much as possible. This should include a consistent and easy to use network of cycle routes. Building new cycle routes or designating cycle lanes is not enough because more needs to be done to adjust the infrastructure, organise cycle traffic and put up clear signage to mark the beginning and end of cycle routes. It is vital to educate and promote safe behaviour. Cyclists are not legally required to hold a driving license. While they are expected to know traffic regulations and follow them, practice shows that this is not always the case.

References

1. European Commission, Communication from the Commission to the European Parliament, the Council, the European Economic and Social Committee and the Committee of the Regions on a roadmap for moving to a competitive low carbon economy in 2050, COM(2011) 112 (Brussels, 2011)
2. European Commission, White Paper "Roadmap to a Single European Transport Area – Towards a competitive and resource efficient transport system". COM(2011) 144 final (Brussels, 2011)
3. European Commission, Green Paper "Towards a new culture for urban mobility", COM (2007) 551 (Brussels, 2007)
4. S.A. Goldsmith, Reasons why bicycling and walking are and are not being used more extensively as travel modes (Federal Highway Administration, 1992)
5. R. Okraszewska, Z. Jagielska, Comparison of Polish and Danish approach to promotion of cycling, (2017)
6. TEMS - The EPOMM Modal Split Tool, available at: http://epomm.eu/tems/index.phtml?Main_ID=2928, (2016)
7. Comprehensive Traffic Research Gdansk (in Polish) (Gdansk Development Office, Gdansk, 2009)
8. Gdansk Traffic Research. A Leaflet (in Polish) (VIA VISTULA, Gdansk, 2016)
9. Comprehensive Traffic Research Cracow (in Polish) (Cracow, 2003)
10. Comprehensive Traffic Research Cracow (in Polish) (Cracow, 2013)
11. Comprehensive Traffic Research Warsaw (in Polish) (Warsaw, 2005)
12. Warsaw Traffic Research (in Polish) (Warsaw, 2015)
13. G. Sierpiński, M. Staniek, I. Celiński, P. Czech, J. Barcik, Identification of pedestrian travel in modeling of modal split and transport accessibility, Logistyka. **4** (2015)
14. Points of measurement of cycling traffic - Bicycle Gdansk' (in Polish), available at: http://rowerowygdansk.pl/start,169,170.html, (2017)
15. The comparison of cyclists measurement in 2011-2014 (in Polish) (Gdansk Development Office, Gdansk, 2014)
16. K. Jamroz, S. Gaca, L. Michalski, M. Kieć, M. Budzyński, L. Gumińska, W. Kustra, T. Mackun, I. Oskarbska, J. Rychlewska, A. Ryś, J. Wachnicka, J. Wierzbicka, Protection of Pedestrians. Guidelines for pedestrian traffic organizers (in Polish) (National Road Safety Council, Gdansk, Warsaw, Cracow, 2014)
17. A. Buczyński, Traffic incidents involving cyclists 2006-2008' (in Polish) (Cracow, Poland, 2009)
18. OECD Statistics, available at: http://stats.oecd.org/, (2017)
19. R. Okraszewska, K. Jamroz, M. Bauer, K. Birr, A. Gobis, Identification of risk factors for collisions involving cyclists based on Gdansk example, (2017)
20. E. Symon, Road accidents in Poland in 2015. Report of General Police Headquarters of Poland (Warsaw, 2016)
21. P.L. Jacobsen, Safety in numbers: more walkers and bicyclists, safer walking and bicycling, Inj. Prev. **9** pp. 205–209 (2003). doi:10.1136/ip.9.3.205
22. L. Michalski, K. Jamroz, K. Grzelec, S. Grulkowski, D. Kaszubowski, R. Okraszewska, K. Birr, W. Kustra, Strategy of Transport and Mobility for Metropolitan Area (in Polish) (Gdansk Metropolitan Area, Gdansk, 2015)

Pedestrian safety management using the risk-based approach

Aleksandra Romanowska[1,*], *Kazimierz* Jamroz[1], and *Wojciech* Kustra[1]

[1]Gdansk University of Technology, Faculty of Civil and Environmental Engineering, Gdansk, Poland

Abstract. The paper presents a concept of a multi-level pedestrian safety management system. Three management levels are distinguished: strategic, tactical and operational. The basis for the proposed approach to pedestrian safety management is a risk-based method. In the approach the elements of behavioural and systemic theories were used, allowing for the development of a formalised and repeatable procedure integrating the phases of risk assessment and response to the hazards of road crashes involving pedestrians. Key to the method are tools supporting pedestrian safety management. According to the risk management approach, the tools can be divided into two groups: tools supporting risk assessment and tools supporting risk response. In the paper attention is paid to selected tools supporting risk assessment, with particular emphasis on the methods for estimating forecasted pedestrian safety measures (at strategic, national and regional level) and identification of particularly dangerous locations in terms of pedestrian safety at tactical (regional and local) and operational level. The proposed pedestrian safety management methods and tools can support road administration in making rational decisions in terms of road safety, safety of road infrastructure, crash elimination measures or reducing the consequences suffered by road users (particularly pedestrians) as a result of road crashes.

1 Introduction

Poland is considered the EU's most dangerous and least friendly country for pedestrians. Figures show that in Poland between 2004-2015 there were 160,000 fatal and injury crashes involving pedestrians in which 17,800 pedestrians were killed. The causes include dangerous driver behaviour (such as: speeding, red light running, not respecting pedestrians' right of way, overtaking at zebra crossings, etc.), dangerous pedestrian behaviour and poor road infrastructure. Pedestrian safety is a recognised problem and one that has been the focus of road safety efforts for a number of years. There are road safety programmes and strategies both at country and regional level in Poland which give pedestrian safety a priority making it one of the main goals of Poland's efforts towards vision zero [1,2]. While Poland has been able to gradually reduce pedestrian fatalities (in 2008-2014 pedestrian deaths decreased by 40%) the results are still far from the expectations.

To ensure that pedestrian safety is handled properly, systemic measures are required designed to identify and assess the hazards and where they occur. This provides the basis for identifying the key problems and proposing effective measures relevant to the problems. The

* Corresponding author: aleksandra.romanowska@pg.edu.pl

pedestrian safety management process should include a number of stages and reflect all phases of a road accident and all contributing factors such as the road user, vehicle, road infrastructure and pedestrian infrastructure and the roadside [3]. To that end extensive knowledge is needed on how the man-vehicle-road system operates putting special emphasis on the pedestrian being the most vulnerable road user.

The paper proposes a pedestrian safety management system for different areas and levels of management. It builds on a risk management method implemented for pedestrian safety management purposes. Special attention was paid to tools that are designed to support the pedestrian safety management system across all levels, with the focus on tools for risk assessment as a key step in pedestrian safety management.

2 Theory

As road transport continues to grow, new challenges emerge for pedestrian safety that call for appropriate response to the increasingly high risk for pedestrians. Over the last few decades road safety approaches have evolved significantly starting from user focus and the 3E approach. The term "safe system" gradually emerged along with beginnings of an interdisciplinary and systemic road safety management [4]. Pedestrians were now viewed as sensitive and vulnerable road users that must be given priority in road traffic. Today to ensure that road safety and its most vulnerable users are given the treatment they deserve we need a comprehensive and multi-sectoral approach taking a multi-faceted view on road crashes, their causes and circumstances suggesting the most effective ways to prevent those crashes from happening in the future [4,5].

A systemic approach to pedestrian safety management was proposed earlier in [1,6]. It is founded on the risk-based method with elements of behavioural [7–9] and systemic [10,11] theories. When applied to road engineering they help to establish a formalised and repeatable process which integrates risk assessment and risk response to the hazard of fatal and injury pedestrian crashes. The concept is presented in Figure 1.

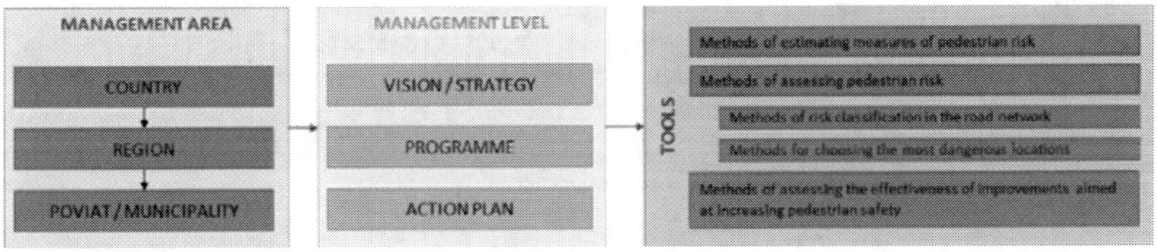

Fig. 1. Risk-based pedestrian safety management concept [6].

The proposed method distinguishes three areas of pedestrian safety management. At the country level, represented by central authorities, key decisions are made: road safety programmes, legislation on safety and decisions that will trickle down to the lower levels. Implementing the decisions is the responsibility of central bodies tasked with coordinating specific issues such as organising the road safety system, road traffic enforcement, road infrastructure, legal solutions, road rescue and education, information and communication. The decisions taken at lower levels (region, county, municipality) take account of national policy. Those responsible for the implementation include road authorities, transport operators or other delivery bodies. Because they have less power, lower level bodies can do less compared to the national level. Despite that, they can manage pedestrian safety in the following areas: road safety structures, road user education, traffic enforcement and control, road infrastructure and rescue.

Whatever the management areas, the actions should be delivered both at the strategic, tactical and operational level [1,3]. At the strategic level political decisions are taken and long-term goals and objectives are formulated that are the basis for pedestrian safety management at the lower levels. The tactical level explains how strategic level goals should be delivered in the mid-term by defining ambitious yet achievable goals and identifying the most effective measures. The operational level implements concrete measures that are undertaken directly by road authorities, town planners, transport operators and road users. Road safety system users are influenced at the particular levels by road safety strategies, programmes and action plans that come with funding systems. To ensure that the influences are effective, support is needed in the form of tools, procedures and methods that help with decision-making and responding to pedestrian safety needs. At this stage the concept moves into the risk-based approach in risk engineering, a formalised and repeatable procedure integrating risk analysis and risk assessment and response to eliminate risk or bring it down to acceptable levels [1,12]. The approach distinguishes two basic steps: risk assessment which includes identifying risk context, sources of hazards and evaluating risk measures; and risk response using a set of tools, methods, procedures and processes that help to decide which types of risk should be avoided, transferred, reduced or accepted [3].

3 Tools for pedestrian safety management

Drawing on the authors' concept of three-level pedestrian safety management using the risk-based method (Fig. 1), the tools proposed in this paper can be used to aid the delivery of risk management steps. These include:

- Risk assessment through: analysis of the problems and selection of risk groups (segmentation) based on historical data, estimation (forecasting) of safety measures (e.g. number of fatalities), risk classification for specific areas or road sections under analysis, road safety audits, inspections and controls, identification of hazardous sites.
- Risk response through: selection of the most effective solutions for pedestrian safety, implementation of the solutions and enforcement, monitoring and evaluation, informing road users, legislations, guidance and recommendations, communicating risk to the public.

Selecting the right and effective tools is important because risk reduction or elimination can only be achieved thanks to a thorough risk assessment. In particular, strategies, programmes and action plans, all of which are elements of pedestrian safety management at the different levels, must build on a detailed risk assessment designed to set goals and targets in priority areas where urgent intervention is needed. This is why the paper focuses on this risk management step. A number of examples of tools can be presented that can be used for risk assessment. Some of them are presented below.

3.1 Methods for estimation of pedestrian safety measures at strategic level

Estimating pedestrian safety measures is possible using numerical models, which helps to study the relation between pedestrian safety measures and variables describing an area in question and its residents. This tool provides the basis for strategic road safety management which should build on long-term forecasts of safety measures and understanding how the activities (investments, legislation, etc.) affect road safety.

At the strategic level risk is assessed using societal risk referring to entire groups of society in a given area. Jamroz [3,13,14] proposed a two-component model used to calculate general societal risk, where the estimated measures are the result of the product of exposure to a specific type of risk and mean consequence of a selected category in a unit of time. Building on the model, efforts can be undertaken to develop models for estimating pedestrian fatalities at country or regional level. Jamroz et al. [1] developed a country-level model (Fig. 2) for estimating pedestrian fatality rate RFR_p on the basis of demographic, geographic, economic, societal and transport parameters that characterise analysed countries in the years of the analysis (a total of 321 country-years were analysed). With its high regression coefficient (> 0.9), the model helps to estimate the country-level pedestrian fatality rate based on the available data.

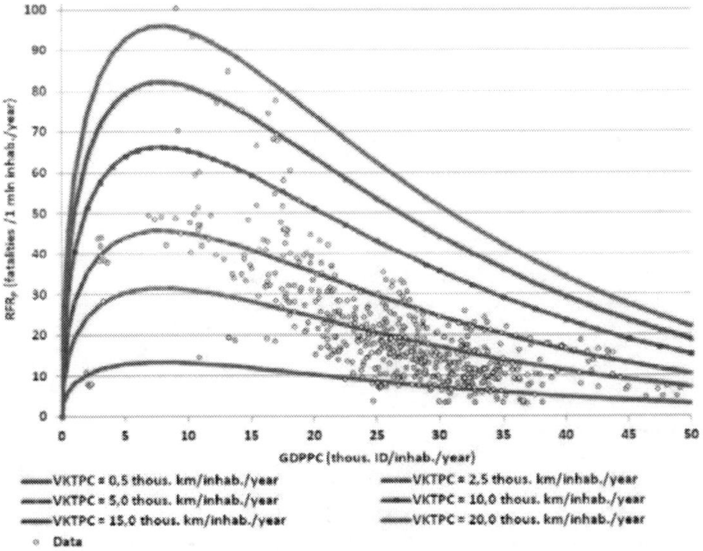

Fig. 2. The fatality rate in relation to the level of socio-economic development modelled at country level [1].

A similar approach to modelling the pedestrian fatality rate was applied when studying Polish sub-regions that are consistent with NUTS-3 territorial units. The collected data included population, areas, gross domestic product, motorization rate and length of paved roads. Analyses were conducted for 66 Polish sub-regions in the years 2009-2012. Two pedestrian safety measures were analysed: pedestrian fatality rate expressed with the number of pedestrian fatalities per population RFR_p and pedestrian fatality rate expressed with the number of pedestrian fatalities per area RFR_a. Further analysis helped to develop a model (1) for RFR_a based on the relations observed in an analysis of correlation. The model's coefficients have a very high level of significance (p < 0.1) and coefficient R^2 was obtained in the order of 0.95. The results are presented in Fig. 3.

$$RFR_a = \alpha_1 * GDP_{pA}^{\alpha 2} * exp(\alpha_3 * GDP_{pA} + \alpha_4 * DOP + \alpha_5 * DCR) \quad (1)$$

where:
- RFR_a – pedestrian fatality rate (fatalities/100 km^2/year),
- *DOP* - population density (population/km^2),
- GDP_{pA} - gross domestic product per area in m PLN/km^2,
- *DCR* - the number of passenger cars per km^2,
- α_1, ..., α_5 - equation coefficients selected to the model through analysis.

In the case of a regional model, due to limited data (only 3 year data on 66 Polish sub-regions) and too few explanatory variables to be analysed (e.g. while vehicle-kilometres travelled would probably be a better predictor variable, they are not easy to obtain at the level of sub-regions) the model should be treated as a point of departure for further research to help identify which factors may affect the pedestrian fatality rate in Polish sub-regions. However, by using a territorial division that is common across the European Union, we can also use data from other countries. The resulting model could be used for forecasting long-term changes in pedestrian fatalities on Polish roads and those in Europe's regions and sub-regions.

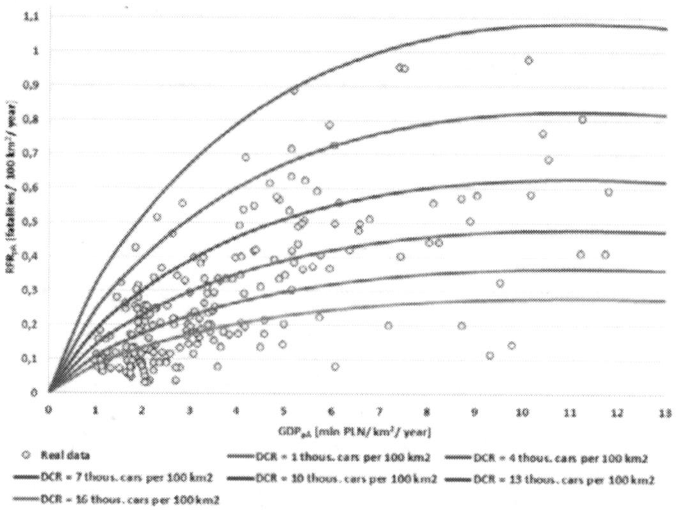

Fig. 3. The fatality rate in relation to the level of socio-economic development modelled at regional level for Polish sub-regions NUTS-3.

Some simple modelling was also conducted for Polish cities that are county capitals. Although the results did not produce satisfactory results, they helped to outline the relations between pedestrian crashes and the variables that are typical of those cities.

3.2 Methods for risk classification at the tactical level

Risk classification should be based on in-depth analyses of pedestrian fatal and injury crashes and risk levels adopted for selected risk measures. Since 2006 Poland has been a member of EuroRAP (European Road Assessment Programme). Building on the programme methodology [15], the risk for pedestrians is classified on a regular basis using crash data. The results are mapped on risk maps available to the public on-line. Risk levels are presented using a clear five band scale of colours: green meaning the lowest risk (and the best safety) with black showing the highest risk and the poorest safety. By using this method risk can be classified at country and region level; we can select areas (region, county), road sections and junctions where the risk for a pedestrian of being involved in a fatal or injury crash is the highest.

To identify pedestrian societal risk, we use pedestrian fatalities and serious injuries as a general risk measure and the demographic rate or density of fatal and serious injury pedestrian crashes as a normalised measure. The analysis is conducted for a 3 year period. This is to avoid the effects of periodical variations or varying road crash numbers due to e.g. the weather (long and snowy winter) and road improvement which may affect the above measures. The boundaries of safety measures are selected based on analyses of areas (regions and counties) and road networks (national, regional and local). The approach used here is probabilistic and engineering (expert) [3]. Risk classification may be done at the regional and county level and for national and regional roads. An example is given in Fig. 4, where the demographic rate of serious accidents is measured as the number of crashes per population and density of serious accidents is measured by comparing the number of crashes to the length of road section.

The method can be used both at the strategic and tactical level of pedestrian safety management. Risk classification of areas and roads can be helpful for authorities (central, regional) when they take decisions at the strategic, tactical or operational level and for road authorities and local authorities when they select pedestrian safety improvement measures and where these should be located. The results of the classification can be helpful with promoting safety assessment, conducting research, planning and developing safety strategies and programmes and implementing effective actions where they are most needed. The method helps select: areas (region, county), long road sections (on national and regional roads), short sections of streets and junctions (in cities) that have the highest risk for pedestrians to become involved in a road fatal or injury crash.

The results of risk classification identify areas/road sections where pedestrian risk is the highest (red and black) and require an intervention to reduce or eliminate the hazard. Following the analyses, risk can be monitored in the years to come to keep track of any changes in pedestrian safety.

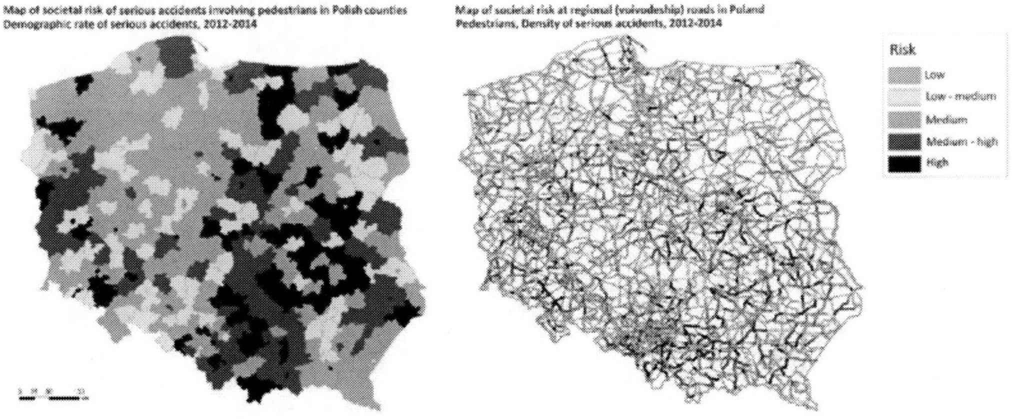

Fig. 4. Pedestrian risk classification based on EuroRAP methodology.

3.3 Methods for identifying hazardous sites at the operational level

Risk assessment at the operational level is supported with identification of hazardous sites. Methods for identifying hazardous sites can be used as a tool for supporting pedestrian safety management. One such method is the "stepping kilometre", presented in [16]. In the method roads are analysed to identify sections where the risk of a road crash is the highest. A one kilometre section is selected on a road with the highest fatal and injury crashes or the highest casualty density. The analysis is conducted in 5 steps: (step 1) determining the first kilometre of a road section 0.0 + 1.0 km and dividing it into 100 m long sections; (step 2) adding up fatal/ injury crashes and casualties for the first ten 100 m long sections; (step 3) shifting by a 100 m and adding up fatal/ injury crashes and casualties on 100 m long sections from 1 to 11, next from 2 to 12, etc., until the final 100 m section; (step 4) summarising the fatal/ injury crashes and casualty numbers for all sections on 1 km; (step 5) analysis of the data will help to select the section with the highest number of fatal/ injury crashes and casualties for the whole road.

Figure 5 shows an example of how the method is used: a road section of 7 km length is analysed according to the above 5-step methodology. The analysis helped to identify three sections (marked on the diagram), where the number of fatal and injury crashes exceeded the critical value, that require an intervention to improve road safety.

 DOI: 10.1051/matecconf/201712201007

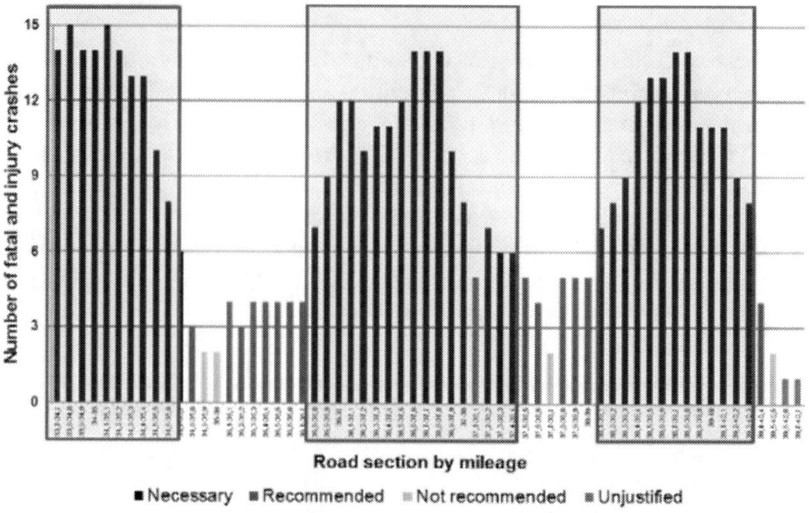

Fig. 5. Identifying hazardous sites with "stepping kilometre" method [16].

The method can be used to identify and assess risk on sites where pedestrian risk is particularly high. This can be the basis for recommendations on measures to be implemented to eliminate or reduce the risk for pedestrians.

The tool can support the pedestrian safety management process at the operational level, in particular. It helps to analyse, asses and classify risk for further risk responses using engineering and expert solutions and by formulating guidelines and recommendations for road authorities.

4 Conclusions

Pedestrian safety in many developing countries continues to be one of the most important and unsolved problems of road safety. This is why efforts in these countries to improve pedestrian safety are vital and will help meet the European Commission's road safety recommendations. Poland is one of the countries where pedestrian safety is a noticeable issue. With no comprehensive approach to pedestrian safety or tried and tested and effective pedestrian safety tools, the country needs a systemic approach to pedestrian safety management across all levels. This is to include an exact identification of pedestrian risks, selecting effective and economically efficient interventions and a systematic monitoring of risk and communicating it. Based on risk, the pedestrian safety management method proposed in the paper offers a tool for a thorough analysis and risk assessment and subsequently a selection of effective risk response measures. The method, however, must be accompanied by tools to support effective and efficient pedestrian safety management. The relevant tools for the strategic, tactical and operational level are presented in the paper. Some are already applied in practice while others require further research.

References

1. K. Jamroz, M. Kieć, W. Kustra, Protection of Pedestrians as the Key Action for Implementing Poland's Vision Zero, TRB 94th Annu. Meet. Compend. Pap. **188** (2014)
2. National Road Safety Programme for 2013-2020 (Polish National Road Safety Council, Warsaw, 2013)

3. K. Jamroz, *Metoda zarządzania ryzykiem w inżynierii drogowej* (Politechnika Gdańska, Gdansk, 2011)
4. A.S. Hakkert, V. Gitelman, Thinking about the history of road safety research: Past achievements and future challenges, Transp. Res. Part F Traffic Psychol. Behav. **25** pp. 137–149 (2014). doi:10.1016/j.trf.2014.02.005
5. Road Safety Guidelines (Multitaleral Development Banks, 2014)
6. A. Romanowska, K. Jamroz, L. Gumińska, Pedestrian road safety management, in: 27th Eur. Conf. Oper. Res., (Glasgow, 2015)
7. G.J.S. Wilde, The Theory of Risk Homeostasis: Implications for Safety and Health, Risk Anal. **2** pp. 209–225 (1982). doi:10.1111/j.1539-6924.1982.tb01384.x
8. R. Naeaetaenen, H. Summala, Road-user behaviour and traffic accidents, Publ. North-Holl. Publ. Co. (1976)
9. R. Fuller, Towards a general theory of driver behaviour., Accid. Anal. Prev. **37** pp. 461–72 (2005). doi:10.1016/j.aap.2004.11.003
10. E. Hollnagel, Mental models and model mentality, in: Tasks, Errors, Ment. Model., pp. 261–268 (Taylor & Francis, Inc., 1988)
11. P. Larsson, S.W.A. Dekker, C. Tingvall, The need for a systems theory approach to road safety, Saf. Sci. **48** pp. 1167–1174 (2010). doi:10.1016/j.ssci.2009.10.006
12. K. Jamroz, S. Gaca, L. Michalski, M. Kieć, M. Budzyński, L. Gumińska, W. Kustra, T. Mackun, I. Oskarbska, J. Rychlewska, A. Ryś, J. Wachnicka, J. Wierzbicka, *Protection of Pedestrians. Guidelines for pedestrian traffic organizers* (in Polish) (National Road Safety Council, Gdansk, Warsaw, Cracow, 2014)
13. K. Jamroz, L. Smolarek, Road safety management tools for country strategic level, in: 16th Road Saf. Four Cont. Conf., (Beijing, 2013)
14. K. Jamroz, The impact of road network structure and mobility on the national traffic fatality rate, in: EWGT 2012. 15th Meet. EURO Work. Gr. Transp. Procedia - Soc. Behav. Sci. 54, pp. 1370–1377 (2012)
15. K. Jamroz, M. Romanowska, Idea and methodology of the EuroRap programme, www.eurorap.pl. (2007)
16. K. Jamroz, S. Gaca, L. Michalski, W. Kustra, J. Zarembski, Multivariate analysis for identyfying hazardous sites recommended for installation of speed control devices and red-light cameras (in Polish) (Gdańsk, 2013)

Methodology for assessing the lighting of pedestrian crossings based on light intensity parameters

Piotr Tomczuk[1,*], *Marcin* Chrzanowicz[2], and *Tomasz* Mackun[3]

[1]Warsaw University of Technology, Faculty of Transport, Warsaw, Poland
[2]Warsaw University of Technology, Faculty of Electrical Engineering, Warsaw, Poland
[3]Gdansk University of Technology, Faculty of Civil and Environmental Engineering, Gdansk, Poland

Abstract. One of the possible preventive measures that could improve safety at crossings is to assess the state of illumination of the lighting installation located in the transition area for pedestrians. The City of Warsaw has undertaken to comprehensively assess the pedestrian crossings to determine the level of road safety and the condition of lighting. The lighting conditions related to pedestrian crossings without traffic lights in three central districts of the city were investigated. The conducted field research and the work of the team of experts lead to the development of tools to assess the level of risk due to the lighting conditions measured at night. The newly developed and used method of assessment and the experience gained should provide a valuable contribution to the development of uniform risk assessment rules for pedestrian crossings in Poland. The authors of this paper have attempted to systematize the description of the method of evaluation of the lighting installed in the area of pedestrian crossings.

1 Introduction

As is apparent from the reports on road safety [1,2] in Poland for many years, there is large number of accidents at pedestrian crossings (Fig. 1).

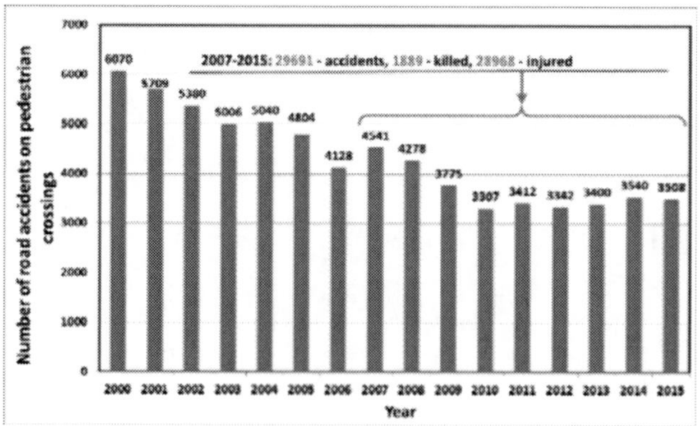

Fig. 1. The number of road accidents on pedestrian crossings in Poland year 2000-2015 [1].

* Corresponding author: p.tomczuk@wt.pw.edu.pl

Action was taken to improve the safety of vulnerable road users in cities [3]. Despite the endeavour to improve the safety of pedestrians in Warsaw in 2011 - 2015, there was no clear trend of any reduction in the number of accidents with pedestrians (434 - 490 accidents / year). Only in 2015 was there a decrease down to 390 accidents. Due to their low statistical numbers, accidents with pedestrians were no longer a map of "black points." To act preventively, a map of the level of hazard for pedestrian crossings should be created, regardless of whether there had been an accident involving pedestrians or not. Acting with the aid of a map of dangers, the road manager can be proactive, not waiting for accidents and fatalities to take place.

The lighting of pedestrian crossings is one of the elements of road safety control [3]. In order to improve the current state of hazards to vulnerable road users a comprehensive inspection of pedestrian crossings must be carried out [4,5]. Recommend the pedestrian crossings to modernize in line with a full assessment of the other road safety factors. On this basis the risk factors should be identified and classified. As a result, it will be possible to mark the pedestrian crossings which require corrective action as regards lighting infrastructure.

It should be noted that uniform and comprehensive safety control procedures, dedicated to pedestrian crossings with a focus on parameterized factors of lighting, have not been developed and implemented in Poland so far. The authors of this publication have made efforts to systematize the process of evaluating the existing lighting of pedestrian crossings for the purposes of the upgrades carried out in Warsaw. It should be noted that the measures taken to evaluate the state of illumination were carried out in parallel with the work of the Road Safety Movement team of auditors, whose task was to assess the other elements affecting the safety of pedestrians.

One of the elements to be evaluated was the lighting of the pedestrian crossings. In urban agglomerations, the assessment of the lighting infrastructure is very important from the point of view of maintaining or improving lighting conditions. The City of Warsaw has undertaken to comprehensively evaluate the existing pedestrian crossings to determine the level of road safety and assess the lighting. Research on pedestrian crossings without traffic lights was carried out in the three central districts of the city. It should be emphasized that no similar project had been carried out in Poland before. The research literature [4–6] preceding the conceptual and field work does not indicate any useful methodologies for the evaluation of lighting parameters. The authors of the literature on the subject encountered just one case [4] where an attempt was made to reach an objective numerical pedestrian risk assessment, giving weight points and weighing the individual parameters of pedestrian crossings. However, in this method the assessment of lighting does not contain the essential quality parameters. Research literature [4,5] indicates that the factor of assessing lighting in the proposed evaluation procedures was treated ambiguously and did not define the lighting parameters and values, or the recommended limits. Balance and weight points cannot be transferred directly from Italian to Polish conditions due to the specificity of the local behaviour of managers and pedestrians and different traffic regulations. Therefore, it had to develop an entirely new, comprehensive evaluation method that allows for the identification of the risk factors for the evaluation of pedestrian crossings. One of them was the state of the lighting of pedestrian crossings. The authors of the Polish methods of assessment decided to split the final evaluation into the lighting parameters and other parameters of road safety (because of the difficulty in determining the gravity of the share of the particular criteria influencing the outcome of the final comprehensive assessment of road safety at pedestrian crossings). Therefore, this article discusses only the assessment of the state of illumination.

2 Assessment of the state of lighting in the area of pedestrian crossings

The lighting of pedestrian crossings in urban areas is often implemented by means of a street lighting installation or using special dedicated luminaires. The lights installed at a pedestrian crossing should also ensure that [7]:
- in the case of the driver - the appropriate conditions to identify the road situation and to see the figure of the pedestrian is in the driver's field of view,
- in the case of the pedestrian - appropriate conditions for the observation of the pedestrian crossing environment and approaching vehicles.

Both in Poland and in other countries [7], pedestrian crossings are zones of conflicting interests (like road junctions etc.), for which there are additional lighting requirements. The guidelines in each country are diverse in terms of the performance and assessment of lighting requirements. In Poland, the basic normative document is the Polish Standard PN-EN 13201-2: 2016-03, Road Lighting, Part 2: Performance requirements [8].

A number of lighting parameters for the assessment of the lighting of pedestrian crossings can be used [6,9–11]. Unequivocal verification of the existing lighting conditions can be carried out using the basic parameters which are the horizontal (E_h) and vertical (E_v) illuminance values [7,11]. The values in these planes clearly describe the sufficient lighting conditions in the study crosswalk. The research on the lighting of pedestrian crossings in Warsaw was carried out with the use of the illumination parameters described in the standard [8] that were adopted in the measurement geometry. Figure 2 shows the basic geometry of pedestrian crossings, indicating the directions of traffic and the points of measurement of light intensity. To study the state of the lighting of pedestrian crossings in field conditions grid points in the planes E_h (plane of the road on which there is a transition together with the expectations for the area - points from 1 to 30) and E_v (the vertical plane passing through the axis of pedestrian crossings that defines lighting figure a pedestrian with the viewing direction associated with the direction of traffic - points 31 to 50) were adopted. The measuring points for measuring the value of E_v located at a height of 1 m above the road surface along the axis of the passage. This measure takes into account the view of a driver approaching a pedestrian crossing in situations where disabled people in wheelchairs, people with short children and baby prams cross the road.

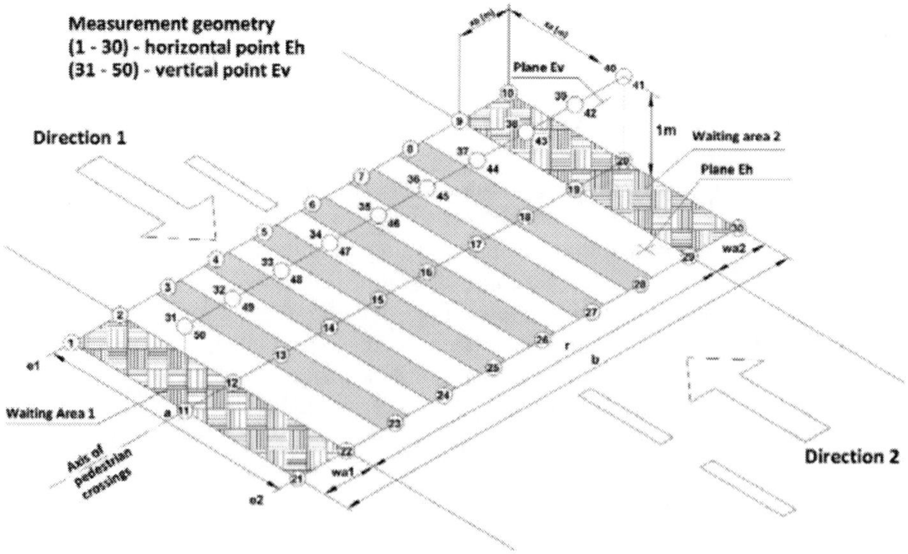

Fig. 2. The basic geometry of pedestrian crossings with the indication of the measurement points.

The standard [8] does not define direct lighting requirements for pedestrian crossings. However, it specifies the requirements for the design and testing of street lighting in the areas of movement on horizontal E_h and vertical E_v. For measurements of the horizontal illuminance E_h at pedestrian crossings the "C" Class of lighting associated with the lighting of conflict zones was adopted (Table. 1).

Tab. 1. C class of lighting and the proposed scale of point marks. Source: [8].

C Class	**Horizontal intensity of illuminance E_h**		**Rating RC**
	E_h w [lx] (lowest value, expected value)	**U_o [lowest value]**	**Points**
C0	50	0.4	6
C1	30		5
C2	20		4
C3	15		3
C4	10		2
C5	7.5		1
no class	< 7.5	-	0

For measurements of the vertical intensity of illuminance E_v at pedestrian crossings we adopted the EV lighting class [8] associated with lighting vertical surfaces (Tab. 2).

Tab. 2. EV class of lighting and the proposed scale of assessments point. Source: [8].

Vertical intensity of illuminance E_v		**Rating $REVD_1$ and $REVD_2$**
EV Class	**$E_{v,min}$ w [lx] (maintained)**	**Points**
EV1	50	6
EV2	30	5
EV3	10	4
EV4	7.5	3
EV5	5	2
EV6	0.5	1
no class	< 0.5	0

The measurements of the state of the lighting on pedestrian crossings in Warsaw were carried out by teams of 3 ÷ 4 that included individuals with experience in conducting specialized field studies of road lighting. All the teams carried out measurements using a single author's procedure. The card of each crossing contains all the data from a visit to the location, photographic documentation, evaluation and the subjective opinion of the evaluation team. The subjective assessment *SE* is issued by the evaluation team during an evaluation in the field and is designed to represent the subjective feelings of the evaluators related to pedestrian crossing zone lighting expectations, illuminating the figure of a pedestrian located on the pedestrian crossing and the condition of the street lighting in the environment where the pedestrians cross. The scale of the scores to describe the subjective *SE* lighting conditions: 0 - very bad 1 - bad lighting, 2 - mediocre, 3 - satisfactory, 4 - good, 5 - very good. The evaluation results are then aggregated in the database of lighting measurements and allow the development of a summary report.

Using the procedure for assigning points for each class C (Table. 1, *RC*) and EV (Tab. 2, $REVD_1$, $REVD_2$) it became possible to designate an objective assessment of the lighting of pedestrian crossings depending on the class of lighting:

$$OE = f_1\, RC + f_2\, REVD_1 + f_3\, REVD_2 \qquad (1)$$

where:
- *OE* is objective evaluation,
- *RC* is evaluation associated with lighting the horizontal plane,
- $REVD_1$ is evaluation associated with lighting the vertical plane in direction 1,
- $REVD_2$ is evaluation associated with lighting the vertical plane in direction 2,
- f_1, f_2, f_3 are the weight factor = 0,33.

The final evaluation of the state of lighting pedestrian crossings *FR* is delivered on the basis of the partial assessment of the subjective and the objective:

$$FR = f_4\, SE + f_5\, OE \qquad (2)$$

where:
- *FR* is final ranking,
- *SE* is subjective evaluation,
- *OE* is objective evaluation,
- f_4, f_5 are the weight factor = 0,5.

In the case of incorrect levels of illumination or improper lighting conditions, technical solutions were proposed to improve the view of pedestrians at night both on the crossing itself and in the waiting area (recommendations included a range of solutions ranging from the cheapest and easiest applications to comprehensive improvements in the lighting system). As a result of the work of established data mart lighting parameters. Pedestrian crossings can be filtered through various criteria, e.g. ratings. Part of the work identified a number of irregularities, e.g. 41 of the examined cases where maintenance procedures on the street lighting should be immediately undertaken.

3 Summary

As a result of our activities, using the research literature and the expertise on the subject, a measuring procedure was developed to unambiguously classify the lighting performance on a test of Warsaw pedestrian crossings. The material prepared this way makes it possible to carry out preventive management measures and investment.

In spite of the existing standards and guidelines, the lighting conditions at pedestrian crossings in Poland are not consistently regulated. There is no clear and verifiable research to define the need for the installation of street lighting and/or additional dedicated lighting. Furthermore there is no legislation defining the lighting conditions that should be met at pedestrian crossings. The technical specifications of tenders for the upgrading of street lighting on pedestrian crossings does not directly define any precise design requirements for contractors. Another problem are the shortcomings at the investment stage and concerning the use of the research verifying the actual state of the lighting on pedestrian crossings. In the future, the development of methods for assessing the state of the lighting will make it possible to continue the work of the monitoring and verification of the solutions to improve lighting conditions.

References

1. Road Safety Annual Report 2016 (OECD, International Transport Forum, 2016)
2. Road accidents in Poland in 2015. Report of General Police Headquarters of Poland (Warsaw, 2016)
3. K. Jamroz, S. Gaca, L. Michalski, M. Kieć, M. Budzyński, L. Gumińska, W. Kustra, T. Mackun, I. Oskarbska, J. Rychlewska, A. Ryś, J. Wachnicka, J. Wierzbicka, Protection of Pedestrians. Guidelines for pedestrian traffic organizers (in Polish) (National Road Safety Council, Gdansk, Warsaw, Cracow, 2014)
4. A. Montella, F. Mauriello, Pedestrian crosswalks safety inspections: safety assessment procedure, in: 4th Int. Symp. Highw. Geom. Des., (TRB, 2010)
5. D. Nabors, M. Gibbs, L. Sandt, S. Rocchi, E. Wilson, M. Lipinski, Pedestrian Road Safety Audit Guidelines and Prompt Lists, FHWA-SA-07-007 (FHWA, U.S. Department of Transportation, 2007)
6. P. Holec, T. Maixner, J. Skála, J. Smetana, J. Tesař, Posouzení přechodových svítidel, Společnost pro rozvoj veřejného osvětlení podzimní setkání, in: (Brno, 2009)
7. P. Tomczuk, Modelowanie, badania eksperymentalne i ocena jakości oświetlenia sylwetki pieszego na przejściu dla pieszych, monografia, seria Transport, zeszyt 91 (Oficyna Wydawnicza Politechniki Warszawskiej, Warsaw, 2013)
8. Polish Standard PN-EN 13201-2: 2016-03 Road lighting Part 2: Operating requirements (n.d.)
9. P. Tomczuk, Assessment model of luminance contrast of pedestrian figure against background on pedestrian crossing, Przegląd Elektrotechniczny. **88** pp. 104–107 (2012)
10. P. Tomczuk, Assessment of the state of pedestrian crossing lighting on the basis of field measurements of luminance, Przegląd Elektrotechniczny. **89** pp. 266–269 (2013)
11. P. Tomczuk, Ocena stanu oświetlenia wybranych przejść dla pieszych na podstawie parametrów natężenia oświetlenia, Przegląd Elektrotechniczny. **90** pp. 303–306 (2014)

MATEC Web of Conferences 122, 02001 (2017) DOI: 10.1051/matecconf/201712202001
GAMBIT 2016

The guidelines and principles for planning and design of road restraint systems

Marcin Budzyński[1,*], and *Marcin* Antoniuk[1]

[1]Gdansk University of Technology, Faculty of Civil and Environmental Engineering, Gdansk, Poland

Abstract. The project RID 3A Road Safety Equipment (RoSE) implemented by the Gdansk University of Technology within the RID programme, aims to conduct a comprehensive study and analyses of different vehicle containment systems (PN-EN 1317) and types of support structures (PN-EN 12767) and their performance. Within the project an analysis will be conducted of available research reports and domestic and foreign experiences looking at road restraint systems and support structures applications. The paper will present a comparison of the guidelines and principles for the design and application of road restraint systems. Similarities and differences in the approach to different solutions will also be discussed. The above analyses will form the basis for the development of tools for selecting and applying road restraint systems in Poland.

1 Introduction

Road safety devices can be divided into two groups. Active devices are designed to handle the impact of out-of-control vehicles, including collisions and crashes. They are specifically designed to minimise the consequences of such events, especially those involving people (injury or death). Passive devices do not come into direct contact with vehicles involved in a crash or accident and are only used to organise and control road traffic, prevent disruptions to traffic and inform motorists and other road users in advance about safety risks or traffic delays. Safety barriers are active road safety devices and are used if the consequences of a crash or accident were greater than those caused by crashing into a barrier (e.g. hitting a tree). To ensure that barriers are effective, they must be designed to successfully handle vehicle impact because their main objective is to protect road users (and roadside users) from fatal injury. The paper presents the experience of other countries in planning and designing road safety devices.

2 Active road safety devices

These devices are primarily used to:
- physically prevent a vehicle from running off the road where this would be dangerous,
- physically prevent a vehicle from crossing into the path of oncoming vehicles,
- prevent a vehicle from colliding with permanent objects or obstacles on the road or roadside,
- secure "special" spots.

* Corresponding author: mbudz@pg.edu.pl

Safety barriers (as the basic active road safety device) should respond to the level risk to traffic safety and roadside hazards. A good design must be based on reliable speed data. Once a hazard that requires treatment is identified, it may have to be removed, moved or mitigated through engineering solutions other than safety barriers. Road safety devices cannot be treated as elements of traffic layout and considered at the last stage of the design when safe solutions can no longer be applied.

3 Overview of Polish guidelines

The document *"Guidelines for safety barriers on national roads"* sets out the requirements for national roads. The guidelines of the General Directorate for National Roads and Motorways (GDDKiA) are mandatory on roads managed by the GDDKiA. Where regional, county and municipal roads are concerned, these requirements are optional [1].

The legal status of the GDDKIA guidelines is a different topic. The term "guidelines" does not exist in Polish legislation. Instead, another term is used, i.e. the national notified application appendix (which means that it is approved by the EU) to a harmonised standard. As a result, guidelines should be presented as an appendix to the EN 1317 standard approved by the EU. This ensures that a specific product meets the requirements and can be used on specific types of roads in a country. The current guidelines, however, can take the form of Director General decisions and will only apply to national roads.

GDDKiA guidelines do not explain how to select the following restraints and their parameters: crash cushions, support structures, anti-glare screens, safety barrier terminals, transitions and temporary safety barriers, pedestrian guardrails, motorcyclist restraints and culvert covers.

To ensure that road and roadside hazards can be assessed objectively, their effect on the safety of road users, third parties and objects is properly understood and the right restraints for the hazards identified are designed, the guidelines distinguish between areas at risk and obstacles. Risk areas are defined by the severity of risk and the distance between the hazard and edge of road.

Safety barriers are selected using diagrams that help to determine the containment level of a particular case. The basic data to be input into the diagram include:

- the severity of the hazard which depends on the distance between the zone protected or obstacle and the road; four levels are identified and the distances are read from nomograms depending on the speed and type of road,
- reference speed – a new term used for the purposes of the guidelines, it is equal to operating speed for road category A, S, GP and Z in accordance with the technical and construction regulations for public roads; for class Z roads it is the lower of two values: allowed speed +10 km/h or design speed +10 km/h, for roads in built-up areas and ramps and interchanges – allowed speed +10 km/h,
- average daily volume of heavy goods vehicles.

To determine barrier containment the designer must answer a number of additional questions. For every barrier adequate distance must be assured from the protected space or area of risk which is to be greater than or equal to the working width. In the case of water reservoirs or slopes, a higher level is acceptable (i.e.W6 rather than W5). As a result, shoulders and bridge plates must be widened, especially because the wheel of a vehicle must not drive off the edge of the structure in case of a crash.

Where a road section includes safety barriers of varying containment levels spaced at short distances, it is recommended to use transition barriers. The containment parameters of these transitions depend on the parameters of the barriers they connect. Unfortunately, the guidelines do not provide precise definitions or examples. As a result, when the designs are prepared by two different organisations, the road/bridge interface may consist of an end of a road barrier and the beginning of a bridge barrier (or the other way round) where it should have a transition section.

The guidelines do not provide any rules for how to select barriers on treated roads, if the width of dividing lanes or the proximity to bridge supports makes it impossible to follow the guidelines. Neither is there anything on how to secure road works.

Another problem is the scope of barrier application. There are no guidelines for the road network except the national roads. The region of Silesia is an exception because it has developed its own set of rules [2].

It should be stressed that the guidelines do not take account of the changes in the revised standard PN-EN 1317 [3,4] (e.g. level of vehicle intrusion VI, dynamic deflection D and levels of containment L1, L2, L3, L4a, L4b) because it was published in September 2010, i.e. 5 months after the Polish guidelines were published.

4 Overview of international guidelines

Following an analysis of guidelines and standards from nearly forty countries, similarities and differences between the countries were identified. Some countries developed their own guidelines while others have adapted existing solutions to varying degrees (Poland's guidelines are an example because they are based on German guidelines) [5]. Figure 1 shows how the main actors (USA, Australia, Germany) have influenced the geographical scope of the rules.

Fig. 1. Distribution of National Guidelines in Other Countries [5].

As we can see, some guidelines / standards dominate in different parts of the world. This goes back to different standards for testing road restraints which are the basis for the guidelines. Standard EN1317 is Europe's VRS test standard (vehicle restraint system), while the US follows the NCHRP350 [6] and MASH [7]. As we can see, American countries have adopted US standards, central Europe uses German standards and northern and western Europe countries have developed their own rules. Despite that, the guidelines have a number of shared solutions such as decision-making processes, tables or charts.

The parameters that determine the type of restraints to be used are divided into thematic groups. Two main categories are distinguished based on the theory of risk: probability and consequences. The probability category includes factors which, if present in a road cross-section, increase the probability of an event. The consequences include factors that increase the severity and consequences of an accident [8,9].

DOI: 10.1051/matecconf/201712202001

The charts in Fig. 2 – Fig. 6 show the percentage distribution of factors that determine the use of safety barriers in the particular countries. The percentages represent the number of countries that consider a specific parameter when designing and installing safety barriers [10–12].

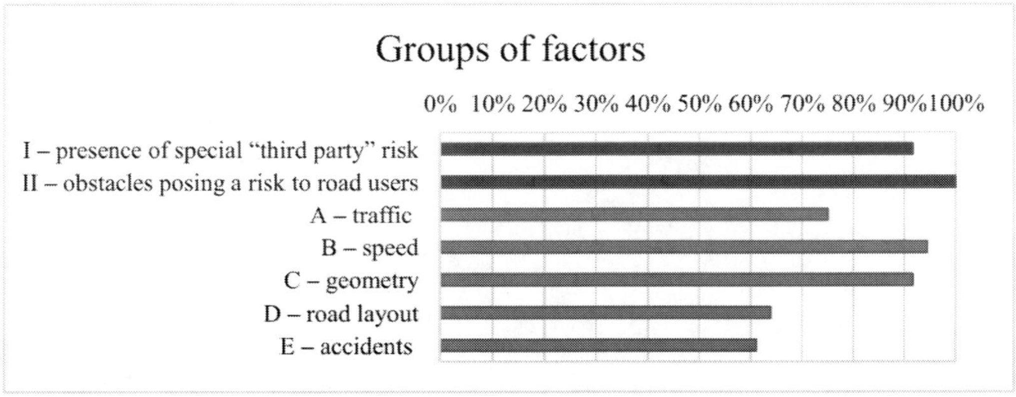

Fig. 2 Groups of factors considered when designing safety barriers based on international guidelines.

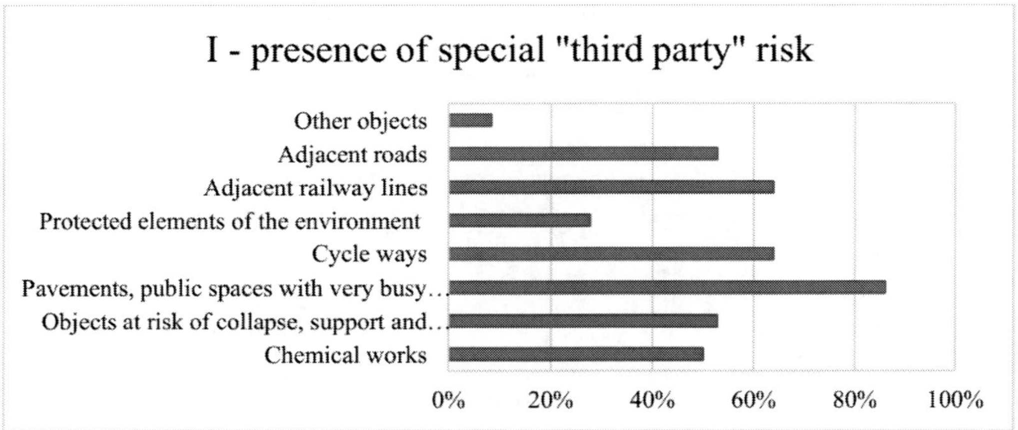

Fig. 3 Groups of factors that pose a hazard to third parties.

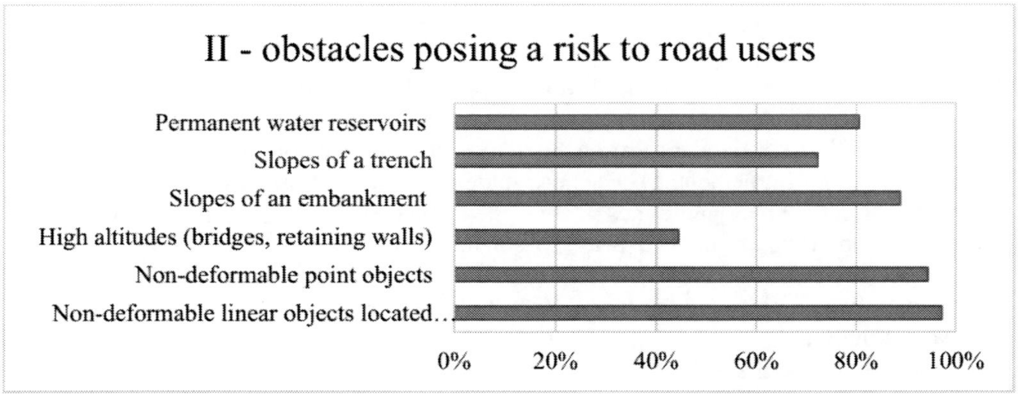

Fig. 4 Groups of factors that pose a hazard to vehicles.

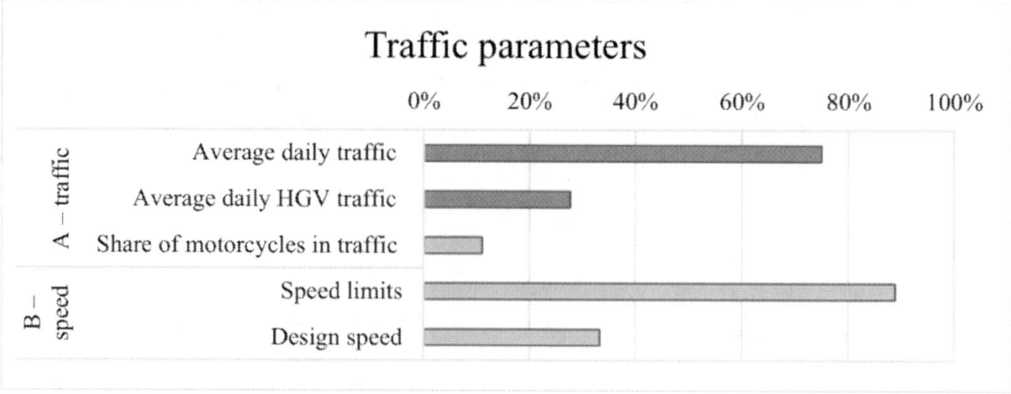

Fig. 5 Traffic conditions for which safety barriers are designed based on international guidelines.

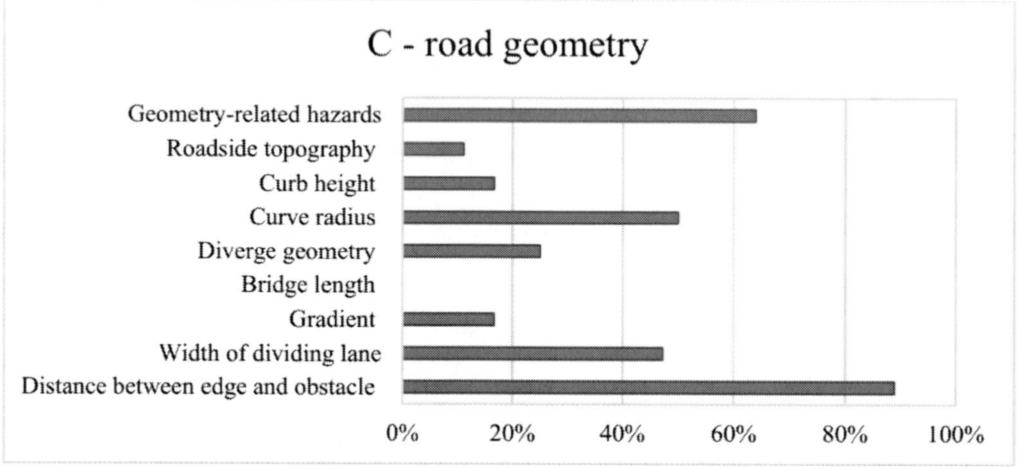

Fig. 6 Geometric conditions for which safety barriers are designed based on international guidelines.

Groups of factors can be identified for how frequently they are used in the analysed guidelines. This can further be used when drafting new guidelines for road restraints system design and application in Poland.

5 Proposed steps

Some early work has been done to develop new guidelines for the design and use of road restraint systems. This was based on a review of guidelines and principles used in other countries, the PN 1317 standard, national roads regulations and comments to 2014 guidelines (which were never implemented). The new guidelines would consist of three parts:

- part one – safety barrier guidelines (for new designs, treated roads, improved roads)
- part two – barriers introduced prior to standard PN EN-1317 and temporary barriers (existing roads, road works)
- part three – appendices (symbols and definitions, examples of calculations, catalogue of selected solutions)

Part one will include the following chapters:

- calculating the safety zone
- criteria for using road safety barriers
- criteria for using bridge barriers

- criteria for using transitions, beginnings and ends, barrier gaps and safety elements other than barriers.

Further research will focus on site tests and numerical tests, their analysis and the development of road safety models taking account of roadsides and road safety devices.

Acknowledgements

This article was written as part the research project RID 3A – Road safety elements (OT3-3A/PG), funded by the National Centre for Research and Development and the General Directorate for National Roads and Motorways (GDDKiA).

References

1. Guidelines for the use of road safety barriers on national roads (GDDKiA, Warsaw, 2010)
2. Design Guidelines for the use of road barriers on regional roads (Road Administration Board in Katowice, 2012)
3. Polish Standard PN-EN 1317-1 Road restraint systems, Part 1: Terminology and general criteria for test methods, (2010)
4. Polish Standard PN-EN 1317-2 Road restraint systems, Part 2: Performance classes, impact test acceptance criteria and test methods for safety barriers (2010)
5. F. La Torre, SAVeRS Selection of Appropriate Vehicle Restraint Systems, (2014)
6. NCHRP 350 Recommended Procedures for the Safety Performance Evaluation of Highway Features (Transportation Research Board, 1992)
7. Manual for Assessing Safety Hardware (American Association of State Highway and Transportation Officials, 2009)
8. M. Budzyński, K. Jamroz, Ł. Jeliński, M. Antoniuk, Why are trees still such a major hazard to drivers in Poland?, Transp. Res. Procedia. **14** (2016)
9. K. Jamroz, M. Budzyński, W. Kustra, L. Michalski, S. Gaca, Tools for road infrastructure safety management - Polish experiences, in: Transp. Res. Procedia, pp. 730–739 (2014). doi:10.1016/j.trpro.2014.10.052
10. Road Safety Barriers - Design Guide Part A (Australian Department of Infrastructure Energy and Resources, n.d.)
11. Road Safety Barriers - Design Guide Part B (Australian Department of Infrastructure Energy and Resources, n.d.)
12. A Strategic Road. Technical report (2015)

MATEC Web of Conferences **122**, 02002 (2017) DOI: 10.1051/matecconf/201712202002
GAMBIT 2016

Solid road environment and its hazards

Marcin Budzyński [1,*], *Tomasz* Radzikowski[2], and *Robert* Zalewski[3]

[1]Gdansk University of Technology, Faculty of Civil and Environmental Engineering, Gdansk, Poland
[2]Regional Road Authority in Olsztyn, Olsztyn, Poland
[3]Regional Police in Olsztyn, Olsztyn, Poland

Abstract. Accidents that involve vehicles departing the road tend to have very high severity, as they often result in the vehicle hitting a permanent obstacle (which can be a tree, a pole, facility pillar, front wall of a culvert or a barrier). In recent years, this type of accident caused approx. 19% of all road deaths in Poland. A particularly high risk can be observed on roads located in northern and western voivodships, with many avenues of trees along the roadside. The paper will present the major road user hazards and their locations, based on the example of detailed analyses conducted for the area of the Warminsko-Mazurskie voivodship in Poland. In the paper, the authors will also present measures that can be applied to improve the safety of road users by improving roads managed by local governments in Warminsko-Mazurskie.

1 Introduction

Run-off-road accidents tend to be very severe because when a vehicle leaves the road, it will often crash into a solid obstacle (tree, pole, supports, front wall of a culvert, barrier). A statistical analysis of the data shows that Poland's main roadside hazard is trees and the severity of vehicles striking a tree in a run-off-road crash. The risks are particularly high in north-west Poland with many of the roads lined up with trees. Because of the existing rural road cross-sections, i.e. having trees directly on road edge followed immediately by drainage ditches, vulnerable road users are prevented from using shoulders and made to use the roadway. With no legal definition of the road safety zone in Polish regulations, attempts to remove roadside trees lead to major conflicts with environmental stakeholders. This is why a compromise should be sought between the safety of road users and protection of the natural environment and the aesthetics of the road experience. Rather than just cut the trees, other road safety measures should be used where possible to treat the hazardous spots by securing trees and obstacles and through speed management. Accidents that are directly related to the road environment fall into the following categories (based on police SEWIK database): hitting a tree, hitting a barrier, hitting a utility pole or sign, vehicle rollover on the shoulder, vehicle rollover on slopes or in ditch.

The main consequence of a roadside hazard is not the likelihood of an accident itself but of its severity [1]. Poland's roadside accident severity is primarily the result of poor design or operation of road infrastructure. This comes as a consequence of a lack of regulations or poorly defined regulations and failure to comply with road safety standards.

* Corresponding author: mbudz@pg.gda.pl

As we know from a number of studies looking at how specific road factors affect safety, the roadside environment and its components (vegetation, shoulders, embankments, drainage ditches, poles, signs, engineering objects, etc.) are very critical [1–5].
Safety barriers are also part of the roadside. While they protect motorists from hitting an obstacle or stop vehicles from leaving the road in the case of steep embankments, they are obstacles themselves, which if poorly designed and built, may pose a serious risk.

2 Analysis of statistical data

Between 2013 and 2015 there were 16,500 accidents related to the roadside (11% of all accidents in that period). The accidents involved 20,700 people injured (16%), including 6,400 seriously injured (16%) and 2,100 people killed (24%). Figure 1 shows the types of roadside-related accidents across the country. As much as 61% of fatalities are caused by hitting a tree. The severity of accidents was analysed for the different types of run-off-road accidents (the number of fatalities per 100 accidents). The following are the results: hitting a barrier – 10, hitting a tree - 23, hitting a sign or utility pole – 9, rollover - 7. As the figures show, run-off-road accidents are clearly most severe when they involve hitting a tree.

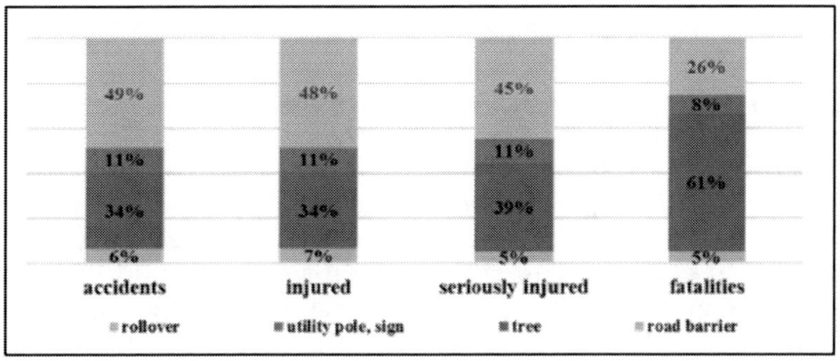

Fig. 1. Types of accidents involving the roadside in Poland.

The region of Warmińsko-Mazurskie was analysed for roadside accidents by road category. Run-off-road accidents are most common on regional roads - 15%, national roads - 9% and others - 10%. As regards fatalities, the majority occurred on other roads at 24%, on regional roads 22% and on national roads 11%. Safety of national roads is much better than in the other road classes. This is because they feature fairly good technical standards and have a high share of upgraded roads (no roadside trees). Figure 2 shows the share of fatalities in roadside-related crashes in the region of Warmińsko-Mazurskie. As much as 62% of casualties occur on county roads and 44% on regional roads.

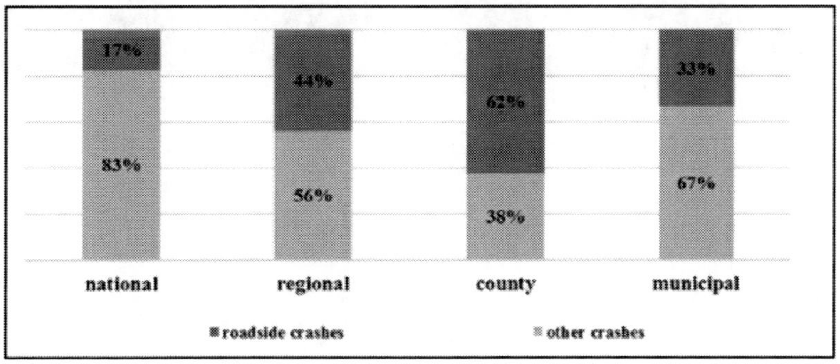

Fig. 2. Share of fatalities in roadside accidents by road category in the region of Warmińsko-Mazurskie.

DOI: 10.1051/matecconf/201712202002

Roadside accidents were also analysed for regional distribution. It was found that in the years 2013 - 2015 the highest share of fatalities was recorded in the regions of Warmińsko-Mazurskie – 38%, Zachodniopomorskie – 34,% and Pomorskie – 34% of all fatalities.

3 Roadside hazard identification

A number of in-the-field tests were conducted looking at road infrastructure and its safety. Based on the findings, a number of elements were identified which present a potential roadside hazard to road users. In 2013 in Poland a road safety inspection method was developed and implemented. The development of the Polish method took account of the experience of other countries [5–7]. Selected sources of hazards were illustrated with photographic documentation (Fig. 3). The sources of Poland's most prevalent roadside hazards include:

- trees close to the edge of the road (up to 3 metres away from the edge of the carriageway the risk is the highest, especially in the area of curves in horizontal alignment, junctions and exits),
- other vegetation obstructing visibility,
- elements of infrastructure which are unyielding (concrete or wooden utility poles, masts, etc.),
- supports of civil engineering objects too close to the edge of the road, unsecured (e.g. bridge supports),
- drainage facilities – vertical concrete front walls of culverts,
- steep embankments,
- poor technical condition of shoulders,
- inadequately terminated, too short, wrong working width and damaged safety barriers.

Photo. 3.1 shows a line of trees along the outer edge of a horizontal curve where the risk of running off the road is much higher than on a straight section or the inner edge of the curve. Photo. 3.2 shows very dangerous safety barrier terminations. Photo. 3.3 identifies a risk in the form of the vertical wall of the culvert. Photo. 3.4 shows the different grade of the carriageway and shoulder, clearly a hazard. All the above examples were identified during a check of Poland's national and regional roads.

Photo. 3.1

Photo. 3.2

Photo. 3.3

Photo. 3.4

Fig. 3. Examples of roadside hazards.

4 Methods for solving the problem

The key to a successful road safety solution lies in an effective road network safety management. A tool proposed in a directive of the European Parliament [8], road safety management fits in with transport safety management in the broad sense. Road network safety management involves a procedure designed to:
- assess safety and identify high-risk sections,
- carry out road safety checks on high-risk sections,
- select the most effective and efficient corrective measures that are appropriate for the funding available,
- communicate the hazards to road users and partners (local authorities, police, partnered businesses),
- monitor safety levels after treatment and evaluate their effectiveness.

Roadside safety management can be delivered at three levels. It begins with the first level which is the strategic level – at the national level. A country's strategic risk in road engineering has to do with its strategic goals such as protecting road users from death or injury, protecting critical infrastructure and strengthening the rescue system. Strategic risk covers the entire system of road transport, including the operation of road infrastructure. Level two is the tactical level and involves regional safety management, i.e. the voivodeship level in Poland. It covers a specific area and identifies the main lines of action. Level three is the operational level and involves concrete objectives at the local level in the area of operating or upgrading a selected road or short road section.

The main factors of hazard at the strategic level that contribute to the severity of run-off-road accidents include: legacy – avenues with roadside trees, speeding because drivers notoriously drive over the speed limit, lack of safe roadside design standards or guidelines, existing infrastructure which "forgives" drivers their mistakes on some sections only.

Subsequent road safety programmes: GAMBIT 2000, GAMBIT 2005, National Roads GAMBIT (sectoral programme for central level road authority) and the National Road Safety Programme 2013 put the emphasis on reducing accident severity as a strategic goal [9–11]. Measures were introduced along with rules for safe roadsides which involved [12]:
- keeping vehicles in lane (signage, narrow hard shoulders),
- use of safety structures (safety barriers, crash terminals),
- upgrading slopes,
- use of safe drainage solutions,
- removing hazardous elements (including trees),
- improving visibility, clarity and consistency of roads.

Within nine years of GAMBIT 2005 implementation, more than 2,200 people were saved from death as a result of removing potential roadside hazards (mainly trees). Sadly, as many as 6,300 people were killed in that period by striking a tree.

The tactical level covers specific parts of a country, mainly its regions. The main risk factors at the tactical level are: the region (the regions in the north and west have the worst track record on all of their road networks), road category (regional roads constituting the biggest problem), length of road sections with trees, element of the carriageway and time of day (high risk during night-time).

Actions implemented at the strategic level (regional scale only) are: construction of new roads, removal of existing trees, use of safety barriers, speed management, public campaigns and primarily the introduction of roadside safety standards.

The "map of interests" is an example of a regional tactical action (Fig. 4). Developed under an initiative of Regional Police in Olsztyn, the map of the primary road network includes national and regional roads. It features sections that are important because of their:

- functional features: identified by road authorities for their function within the road network and technical class,
- transport needs: proposed by haulier organisations as being important for transport and the regional economy,
- landscape features: outstanding avenues identified by environmental services in a stock taking.

Once the risk of hitting a tree was plotted, the result is a map which identifies clearly the priorities (environment or economy) and hot spots that feature conflict of interest. The final safety management level is the operational level. It covers concrete solutions to be applied locally on selected sections of roads or a smaller administrative area of a county or municipality. The factors which increase risk at this level include all road network deficiencies that exist on a given section (defined in Chapter 3). We can also add: road narrowing (wrong-way driving), restricted visibility at junctions and exits, no space for pedestrian traffic and damaged road infrastructure.

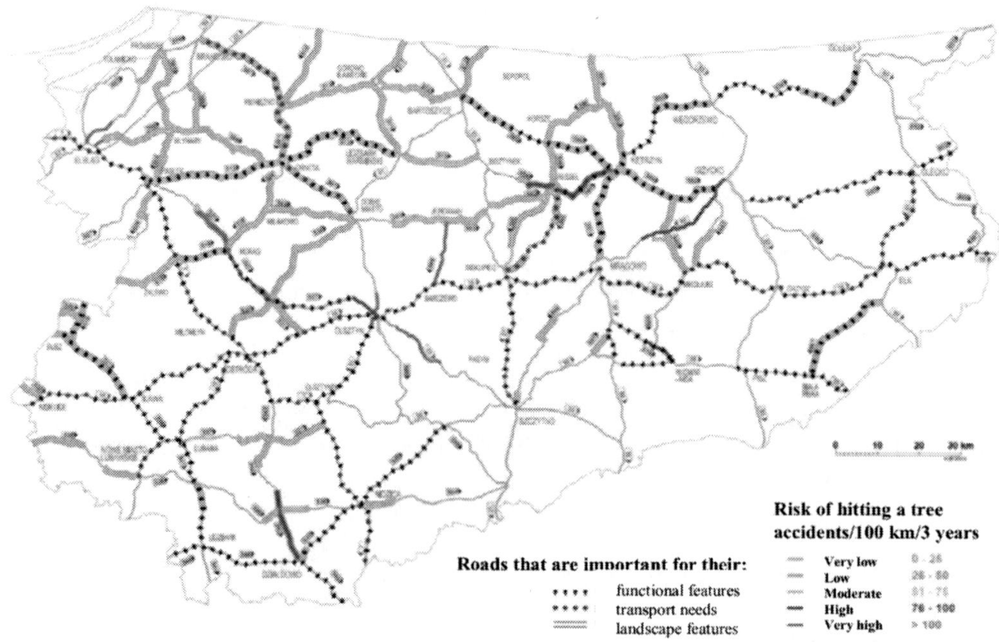

Fig. 4. Map of "interests" in the region of Warmińsko-Mazurskie.

DOI: 10.1051/matecconf/201712202002

A major hindrance at this level is that it only has a very small budget to spend on prevention and treatment. The majority of measures involve signage of trees and sections with trees, speed limits and tree removal. As an example, recommended speed is introduced on sections with trees that are close to the road in the region of Warmińsko-Mazurskie (Fig. 5).

Fig. 5. Recommended speed on sections with trees close to the edge.

5 Conclusions

Over the last twenty five years more than 20 000 people were killed on Polish roads in run-off-road crashes (of which a clear majority involved hitting a tree). Analyses and studies of roadside hazards offer the following conclusions:

1. The main factors that influence the risk of being involved in such a crash are: legacy conditions, road class, length and element of carriageway, hazardous elements at the edge of carriageway (mainly trees), safety measures in place or lack of safety measures.
2. Roadside risk is not the same across Poland. The risk is the highest in the north and west of Poland considering the entire road network; county and regional roads are most at risk among road categories.
3. With no regulations, design standards or cooperation with environmental organisations and institutions, it is very difficult to achieve safety standards and protect road users' lives.
4. To improve roadside safety we must:
   - identify the hazards on the road network, conduct checks
   - conduct research (build models of the effects of selected factors on road safety, effectiveness evaluation)
   - implement safety standards
   - develop guidance and principles for safe roadsides
   - ensure that there is collaboration between designers, road authorities and environmental organisations and institutions
   - exchange experience with other countries.

For years roadside environments have been one of the most neglected aspects of road safety efforts in Poland. Clarity is needed on the effects of roadsides on road safety. We must understand the hazards roadsides cause and implement effective solutions.

Acknowledgements

This article was written as part the research project RID 3A – Road safety elements (OT3-3A/PG), funded by the National Centre for Research and Development and the General Directorate for National Roads and Motorways (GDDKiA).

References

1. M. Budzyński, K. Jamroz, Ł. Jeliński, M. Antoniuk, Why are trees still such a major hazard to drivers in Poland?, Transp. Res. Procedia. **14** (2016)
2. J. Lee, F. Mannering, Analysis of roadside accident frequency and severity and roadside safety management, Research Project T9903 (Washington State Transportation Center (TRAC), Washington, 1999)
3. J.G. Viner, Rollovers on sideslopes and ditches, Accid. Anal. Prev. **27** pp. 483–491 (1995)
4. C. V Zegeer, F.M. Council, Safety relationships associated with cross-sectional roadway elements, Transp. Res. Rec. **1512** pp. 29–36 (1995)
5. S. Cafiso, G. La Cava, A. Montella, G. Pappalardo, Operative procedures for safety inspections on two – lane rural roads, in: IASP, (Catania, Italy, 2006)
6. Road safety inspection guideline for safety checks of existing roads (World Road Association PIARC, France, 2012)
7. Road Safety Inspection RVS 02.02.34 (Bundesministerium für Verkehr, Innovation und Technologie, 2012)
8. Directive 2008/96/EC of the European Parliament and of the Council, On road infrastructure safety management (Brussels, 2008)
9. Gdansk University of Technology, Cracow University of Technology, Motor Transport Institute, GAMBIT 2005 – National Road Safety Programme (Gdansk, Cracow, Warsaw, 2005)
10. Foundation for the Development of Civil Engineering, GAMBIT National Roads Road Safety Programme (Gdansk, 2007)
11. Ministry of Infrastructure and Development, National Road Safety Programme (Poland, 2013)
12. ROSEBUD: Measure against collisions with trees RN134 (landes), WP4 Case Report (Austria, 2005)

Methods of estimating the cost of traffic safety equipment's life cycle

Beata Grzyl[1], *Adam* Kristowski[1], *Kazimierz* Jamroz[1], *Anna* Gobis[1]

[1]Gdansk University of Technology, Faculty of Civil and Environmental Engineering, Gdansk, Poland

Abstract. In the article, the authors discuss the preliminary information necessary to determine the scope and direction of further research conducted within the project called "The influence of time and operating conditions on the durability and functionality of road safety elements". The main objective of the project is to develop the concept of a method for optimizing the life cycle costs of road safety devices. The authors draw attention to the close connection between the decisions taken at the design stage and expenses incurred in the course of maintenance and the use of road safety devices, present the specificity of road infrastructure in terms of life cycle costs, discuss the components of the costs and give examples of the LCC analysis applied to the concrete barrier.

1 Introduction

One of the key priorities of the EU's Europe 2020 strategy is sustainable development involving support for a resource-efficient, environmentally friendly and competitive economy. Many EU documents promoting sustainable development quote the Life Cycle Cost estimation as a basis for decision-making (LCC) [1]. It involves calculating the total cost of the product, item, facility and service, generated in the period from the acquisition of raw materials to waste management, taking into account the phases of design, installation, operation, maintenance, recycling or disposal [2].

2 The purpose of research

The cost of installation of road safety elements, their resilience and reliability depend on many factors, including: detailed technical design, quality and correct installation and the scope and capacity for maintenance and operation. To ensure that spending is rational, in terms of the life cycle cost of a road safety device, the above issues must be considered at the stage of planning and drafting. In many cases, the investor/purchaser analyses only the initial effort (assembling the device) and the technical parameters (expressed with better road safety and driving comfort) and ignores the costs of operation. It should be noted, however, that lower costs borne by the investor/purchaser at the stage of investment, in the long term, are likely to generate significant expenditure due to ongoing operation and maintenance of road safety devices.

* Corresponding author: beata.grzyl@wilis.pg.gda.pl

In the course of the research, identification and analysis will be conducted of major factors generating significant costs, which may be significantly reduced (at the stage of operation and maintenance). The work will outline the relationship between selected factors (e.g. kind of safety barrier, type of road, speed limits, type of collision, traffic, weather conditions, time of year), the scope of the damage and the number of repairs of road safety equipment and the cost of repair, renovations or replacement. Depending on the scope of the input data (e.g. in the form of historical data) in the LCC analysis, account will be taken of the average cost and time to repair and replace the road safety equipment including the impact of various factors (type of barrier, road type, type and scope of collision, frequency and extent of damage, the scope of conducted repair and replacement) as well as road users and social and economic cost.

3 The specificity of road infrastructure and road safety devices

Factors that have a decisive impact on the life cycle cost of road safety devices include: the type and nature of their work, traffic, weather conditions (climatic factors), the location and level of functionality. The fundamental problems associated with the operation of road safety devices which also have an impact on their life cycle cost include [3,4]:

- lack of qualifications and expertise of the contractors – regarding the specifics of road safety devices which results in errors in the installation, repair, reconstruction,
- failure to include in tendering procedures the need for uniform and compatible solutions (types) of devices on the adjacent road sections,
- difficulties with effective enforcement of the warranty of contract performance from the companies managing the specific road section,
- companies responsible for maintaining the road section lack detailed knowledge on the conditions of operation, maintenance, repair, restoration of road safety devices,
- no possibility of a quick exchange of a barrier for another of the same type, as a result other spare parts are used (i.e. from other manufacturers) which increases the probability of the barrier losing its functional parameters.

The main expenditure of the life cycle and road infrastructure costs are borne by road managers. This is related to maintenance and operation (often estimated over a specific period of, e.g. 25 years).

In order to determine the optimal life cycle cost of road infrastructure, costs should be estimated which are incurred by the individual user (e.g. disruption caused by maintenance works, removal of consequences of road collisions and accidents), social losses (e.g. the deterioration of the health and quality of life of local residents) or the impact on the environment (air pollution and noise) [5].

Many road managers in the world in order to increase the efficiency of the expenditure and reduce the costs of maintenance and use of roads and infrastructure and to maximize socio-economic benefits, have developed their own models of LCC analysis. There are different types of models - from simple models taking into account only the cost of supplies, to more complex involving social costs. One of the methods for assessing the design of road construction by using analysis of the annual cost is given in [6]. The publication highlights the difficulties associated with determining the costs of management of road infrastructure and the lack of knowledge of the designers on the scope of maintenance.

In [7] models are analysed for calculating life-cycle costs developed by road managers: COMPARE (United Kingdom), QUEWZ (Australia), Whole Life Costing System (USA) and Highway Design and Management (HDM I-IV). These models, however, are used primarily in the design and construction of roads and pavements. The authors presenting these models [3,4,6–8] pay attention to restrictions in calculating the life cycle costs. The functionality management of road safety devices in terms of the cost of their life cycle. The usability (usefulness) of road traffic safety equipment is expressed mainly with maintaining the continuity of its operations. This includes ensuring that road users can rely on the relevant quality parameters of the transmission of information and vehicle safety, etc. over a specific period. The functionality management of the road safety equipment is based on maintaining a high degree of its readiness for operation, as well as rapid removal of damage or failure [8].

4 The components of the life cycle costs of road safety devices

Device manufacturers use LCC road safety analysis in order to optimize the cost of the life cycle, taking into account safety, degree of functional failure, frequency of maintenance, etc. In practice, however, in many cases the costs incurred in the full life cycle of the device are omitted, and investment decisions are made based on short-term criteria. This causes the negative consequences of financial but also environmental nature (the need for early disposal of waste). Due to the fact that the resilience of some road safety equipment dates back 15-25-30 years, and the costs associated with their operation and maintenance far outweigh the initial outlay, the criterion of the lowest price should not be the sole determinant in making an investment decision. The costs generated in subsequent stages of the life cycle of the road safety device include:
- the concept and definition (includes the cost of market research, analysis, concept, defining the requirements for the product),
- the design and development (includes the cost of the documentation design, prototype production, software, quality management),
- production (includes manufacturing costs, delivery to the market - loading, transport),
- installation - assembly,
- the use and handling (includes the cost of repairs, maintenance, spare parts, technical support, incurred over a given time of the device's operation),
- decommissioning (includes the cost of dismantling, recycling or disposal).

The total costs incurred in the life cycle of a road safety device can also be divided into [8] the cost of acquisition and ownership. The acquisition costs primarily include costs of investment - the purchase of equipment (i.e. the design and manufacturing) and its installation [9]. Ownership costs include operating costs, maintenance, scheduled maintenance, troubleshooting, but also the environmental, social and decommissioning costs. Typically, the cost of ownership outweigh the costs of acquisition.

In relation to the total cost incurred during the life cycle of road safety devices, another possible division is into [8]: user and manager costs. The costs incurred by managers include fees associated with the maintenance and use of road safety equipment: repair, replacement, maintenance of individual items, interventions. In practice, a big problem is determining the value of a random occurrence of events (accidents, collisions) and frequent lack of information about the events. User costs (also included in the cost of ownership) will be borne by an individual or group of road users and include losses resulting from the occurrence of road incidents. These include the costs of incidents, accidents, injuries and fatalities, but also the cost of wasted time of road users, repair of motor vehicles and environment protection [8,10,11].

The essential factors to be taken into account when determining the costs of maintenance and use of road safety equipment include [8,10,11]: the device type, period (winter, summer) in which damage/destruction of barrier occur, road conditions, location of the barrier, road type, current speed limit on the road, the distance between the barriers and the edge of the lane, number of lanes on the road. The amount of the costs associated with the repair also determines the scope of repair, the extent of the damage caused by the impact of the vehicle and the amount of granted compensation.

5 The procedure for conducting the LCC analysis

The process of LCC analysis can be divided into several key stages. An important element of estimating the life cycle costs of road safety devices is to define the actions that have an impact on the reduction or increase in costs and to establish the correlation between them. An important aspect of the analysis are cost data collected from road managers and those responsible for maintenance of road (when such information is collected). A significant element of the LCC analysis is also relevant to the lifetime of the road safety device (15-25-30 years) and the amount of the discount rate. Formulated in this way, input data can be processed using a specific calculation model.

5.1 Methods of LCC analysis. An example of life-cycle cost of a road barrier

There are two main groups of LCC analysis methods [1]:
- simple, uncomplicated comparisons designed to select the optimal variant of the process without discounting,
- complex, including discounted cash flow analysis for the period from installation to withdrawal from use, integrating the different elements of the costs incurred during the life cycle (such as maintenance, operation, repair, dismantling).

The selection of the appropriate LCC method depends on the nature, scope and complexity of the project. Methods that can be used to estimate the components of the life cycle cost of road safety devices are [1]:
- the engineering method to estimate the cost (direct testing of the product, component after component),
- the method of estimating the cost by analogy (an estimate based on experience gained from similar products or technologies),
- parametric cost estimation method (the use of parameters and variables to develop according to estimating cost).

The authors present a sample calculation cost of concrete barriers life-cycle. Calculations performed by estimating the cost by analogy using data and information are contained in [12]. In order to perform an LCC analysis assumptions are taken as shown in Table 1.

Table 1. Assumptions used for the LCC analysis. source: own study based on [12].

Data	Value
Type of road safety device	The concrete barrier prefabricated outside the place of assembly
The average annual cost of use and maintenance (including repairs and routine maintenance)	0.62 PLN/m/ year
The cost of barrier installation	298.14 PLN/m gross [13]
The assumed discount rate	5%
Adopted lifetime	15 years
The total LCC life cycle cost	4273.29 PLN

This LCC analysis indicates that the cost of operation and maintenance of concrete barriers within the given period and adopted output data is more than 14 times higher than the initial effort.

5.2 Restrictions on the LCC application

LCC analysis provides a new approach more widely used, for example, in the process of designing and making an investment decision on the choice of road safety devices with certain characteristics. In practice, however, there are difficulties and constraints characteristic of many techniques and utility costs. The most important ones are presented by the authors later in the article.

1. The LCC analysis method is not rigorous, as a result it can generate different results. This is due to the fact that the costs included in the analysis are only approximate (especially those assigned to the phase of the operation, maintenance and decommissioning). The analysis requires a lot of input data (e.g. difficult to obtain from road managers and maintenance companies).

2. In practice, designers lack knowledge on the technology of repairs to road safety devices and the associated costs. In addition, estimating the costs at the social, administrative and economic design stage is a difficult task.

3. There are no reliable data on the maintenance and use of road safety devices due to the fact that the majority of road managers do not have a suitable method for systematic data collection. There are no accurate models of road structure aging to take account of progress of civilization (including increasing annual volumes of traffic) [11] or to consider changes in the cost calculation.

6 Conclusions

The analysis of the literature and the authors' own experience justify the conclusions, statements and recommendations stated below.

1. The use of LCC at the concept stage, when defining and designing road safety devices is a great opportunity to reduce costs over the life cycle. Due to the lack of information regarding costs of operation and liquidation of a given type of road safety device, further research conducted by the authors includes creating a database of road maintenance costs on the basis of information obtained from road managers and entities responsible for carrying out maintenance work on roads.

2. As indicated in practice one of the reasons for increasing the operating costs of road safety devices is that road managers and maintenance companies apply the old technology and incompatible solutions. If incorrectly selected, the technology increases the cost of repairs and traffic congestion. The rational solution is to use a technology that offers a low replacement cost. It is therefore necessary to analyse the use of barriers with different initial outlays and the cost of repair and maintenance.

3. Despite numerous constraints (such as determining the individual components of costs requires conducting field, laboratory and simulation research), LCC analysis can be applied to assess the effectiveness of variants of the purchase of new equipment or upgrading existing road safety. It provides transparent information about the possible consequences of solutions, including, among others, cost, lifetime and reliability of the road safety device parameter.

Acknowledgements

Research conducted under project implemented within RID programme, funded by the Polish National Centre for Research and Development and the General Directorate for National Roads and Motorways in Poland, for the years 2016-2019.

References

1. PN-EN 63300-3-3 Zarządzanie niezawodnością część 3-3: Przewodnik zastosowań. Szacowanie kosztu cyklu życia (2006)
2. Directive 2009/125/EC of the European Parliament and of the Council of 21 October 2009 establishing a framework for the setting of ecodesign requirements for energy-related products (2009)
3. Bagiński G., Problemy utrzymania drogowych środków brd, in: XI Int. Road Saf. Semin. GAMBIT 2016, (Gdansk, 2016)
4. K. Jamroz, Method of risk management in highway engineering. Metoda zarządzania ryzykiem w inżynierii drogowej (Gdansk University of Technology, Gdansk, 2011)
5. Mainline – Maintenance, renewal and improvement of rail transport infrastructure to reduce economic and environmental impacts, 7 FP EU (Brussels, n.d.)
6. U. Olsson, Projektering med hänsyn till drift och underhåll. Infrastructure design with regard to maintenance and operation. Report No.1983:25 (Construction Research Council, Stockholm, 1983)
7. A. Huvstig, "Whole Life Cost," in: 8th Int. Symp. Concr. Roads, Lisbon, Sept. 13-16, Proceedings, Theme III, pp. 43–51 (1998)
8. H. Karim, Improved Road Design for Future Maintenance – Analysis of Road Barrier Repair Costs, Royal Institute of Technology, Stockholm, 2008
9. B. Grzyl, A. Kristowski, A calculation proposal of labour time input when concreting in difficult atmospheric conditions, Czas. Tech. Bud. Zesz. 2-B. (2014)
10. H. Karim, Road Design for Future Maintenance - Life-cycle Cost Analyses for Road Barriers, (2011)
11. H. Karim, R. Magnusson, K. Natanaelsson, Life-Cycle Cost Analyses for Road Barriers, J. Transp. Eng. **138** pp. 830–851 (2012)
12. S.A. Cooner, Y.K. Rathod, D.C. Alberson, R.P. Bligh, S.E. Ranft, D. Sun, Performance Evaluation Of Cable Median Barrier Systems In Texas. Report No. FHWA/TX-09/0-5609-1. Available: http://tti.tamu.edu/documents/0-5609-1.pdf (2009)
13. www.drogowe.com.pl, (n.d.)

MATEC Web of Conferences **122**, 02004 (2017) DOI: 10.1051/matecconf/201712202004
GAMBIT 2016

Simulation of the impact of dangerous road incidents on driver's loss of time on express roads

Barbara Strzebrakowska[1,*]

[1]Gdansk University of Technology, Faculty of Civil and Environmental Engineering, Gdansk, Poland

Abstract. Incidents on roads are an important factor in traffic flow disruption and a consequent loss of time for drivers. Disruptions can be the result of unexpected road incidents or planned repair works after an incident. These situations differ in the availability of information about the incidents and the resulting rerouting undertaken by drivers. The first part of the paper presents a literature review on the impact of road incidents on traffic and driver's loss of time modelling. Statistical analysis of the database of road incidents on motorways and express roads in the region of Pomorskie was conducted. The average time of arrival of emergency services to a scene and the average time of the rescue action were determined. Simulation research on road incidents was performed for different scenarios with modifications of traffic volume, duration of the road incidents and the degree of road section capacity limitations in various options. The simulation was performed by PTV VISUM software, with the DUE (Dynamic User Equilibrium) module for dynamic traffic modelling. The result of the research is the estimation of the average loss of time for the different variants of the incident, which enabled building a model of the driver's loss of time for specified event parameters.

1 Introduction

On their daily commute, road users choose their means of transport, the route and the time so that the journey is the most efficient in terms of duration, convenience and expense. With time drivers use the best route variant, which results in the traffic reaching equilibrium meaning that no other decision would result in a lower cost for drivers' journey [1]. The network is in a condition of equilibrium as a typical state, that contrary to appearances is very rare. Changes in traffic conditions occur frequently, caused by planned events such as holidays or mass events. The relevant information can be distributed by different methods so that drivers have a chance to change the route, means of transport or decide not to go. In the event of an unexpected incident such as an accident or collision it is impossible for drivers to change the route before setting off. Such events are unexpected and impossible to plan for [2].

* Corresponding author: barstrze@pg.gda.pl

DOI: 10.1051/matecconf/201712202004

2 Existing conditions

2.1 The impact of road incidents on traffic

Road incidents can result in obstruction to normal passage on a section, interchange or junction, which involves a change in the parameters of traffic on the network. The character of this phenomenon is dynamic. All resulting actions and effects of an incident affect relevant interconnected parameters such as [3], section capacity and average speed of vehicles.

The decrease in capacity associated with the traffic incident depends on the number of lanes on the road and number of lanes blocked as a result. The study conducted in 2002 [4] suggests, for example, that a block of one lane on a two-lane road in each direction causes a capacity reduction of about 68% in the direction in which there is a traffic incident. Table 1 shows the detailed results.

Table 1. Percentage capacity of the section during the road incident [4]

Type of restriction	**Capacity during incident (%)**				
	Lane number – express road				
	1 lane	2 lanes	3 lanes	4 lanes	5 lanes
Vehicles in emergency lane	0.45	0.75	0.84	0.89	0.93
1 blocked lane	0.00	0.32	0.53	0.56	0.75
2 blocked lanes	-	0.00	0.22	0.34	0.50

The main effect of reduced capacity is the loss of time of drivers who are indirectly affected by the incident. Loss of time is understood as the difference between total travel time during normal average conditions and total travel time during the incident until normal conditions are restored. The analysis looks at the same time, duration and area for both situations.

2.2 Driver time loss modelling methods

There are several methods to estimate the time loss for road users when an incident occurs on a developed road network. Time loss may be estimated for a particular case referring to the road network, traffic volume, incident duration and type. Macroscopic, mesoscopic and microscopic computer simulations can be used for the calculation. For the analysis of road users' time losses as a result of road accidents each of these methods can be used [5].

For example, time loss estimates with the use of macrosimulation means introducing relevant restrictions on properly calibrated network. These changes will contribute to a change in the distribution of traffic on the network and make it possible to determine the changes in traffic volume in each section. In addition, it will be possible to calculate the loss of time for road users who will benefit from an alternative route bypassing the section where the incident took place.

3 Statistical analysis of road incidents

Databases used in the paper relate to traffic incidents that occurred in the years 2012 - 2015 on the motorway and express roads in Pomorskie. This database was made available by the Regional State Fire Service in Gdansk. In total, in the analysed period there were 177 road incidents on road S6; 55 incidents on the S7 and 166 on the A1 motorway [6]. The most frequent arrival time was 6 to 10 minutes. While it seems quick, it still exceeds 5 minutes. On a large number of routes, it took from 11 to 15 minutes. It was verified that the type of traffic incident did not affect significantly the average arrival time to the incident scene [6]. Response arrival times to the scene of incidents are summarised in the graph in Figure 1.

The time to deal with the consequences of an incident differs a lot. Emergency services, in most cases, took 10 to 100 minutes. There were incidents in which the consequences were dealt with for over 3 h. The time of incident resolution on the S7 did not take more than 1 h 30 minutes. In most cases, it took 20 to 40 minutes. Only 10% of the incidents required actions longer than an hour. Time of dealing with the consequences of traffic incidents on the A1 motorway in most cases lasted from 10 minutes to 2 hours. As many as 8% of incidents required interventions of firefighters lasting over 2 hours. On average firefighters' actions took about 30 minutes [6].

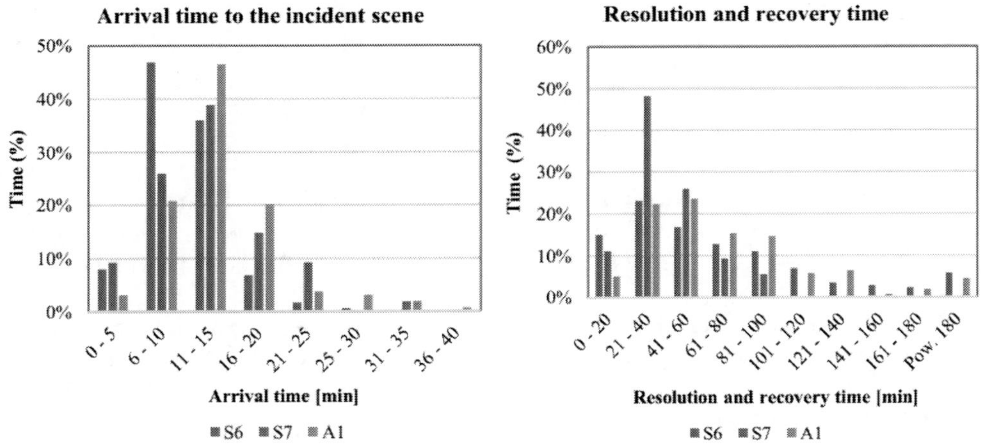

Fig. 1. Time of response arrival and resolving consequences of the incident (own study based on [6]).

4 Simulation studies

4.1 Assumptions and variants of analysis

To create a model of time loss caused by traffic incidents the analysis was performed for a section of the highway, with the parameters adopted as in the model for the Metropolitan Area of Gdansk, Gdynia and Sopot in which link capacities and section resistance function are based on measurements performed by University of Technology employees. The simulations were conducted for various parameters, therefore the variants provided, differed within such parameters and values as in Table 2.

Tab. 2. The values of the parameters included in the model (own study).

Parameter	Analysed scenarios
Traffic volume	500, 1000, …, 3500, 4000, 4400 [P/h]
Incident duration	10 min, 20, 40, …, 160, 180 [min]
Level of capacity reduction	25%, 68%, 98% [4]

4.2 The results of simulation studies

The driver's time loss was obtained after conducting the simulation for all the variants of traffic volume and duration of the traffic incident for a blocked emergency lane. The resulting data were mapped in the graphs presented in Figure 5. The graphs show loss of time depending on the duration of the incident (the X axis) for different traffic volumes (the individual curves of the graph).

A traffic incident in which a vehicle or vehicles occupy an emergency lane results in speed reduction and limited lane capacity (25%). Although traffic movement is still possible on two lanes, this movement is hampered. For traffic below 3000 veh/h time loss is insignificant: 0 - 10 veh/h in total for all drivers. If the traffic volume is close to the section's capacity, time loss values begin to grow rapidly. This is due to the fact that the queue which was formed, started dispersing slowly. This applies to a traffic volume with a capacity of over 80%.

A traffic incident, in which one lane is blocked takes place when it is not possible to remove vehicles from the road onto the shoulder or emergency lane, or when it is possible to remove the vehicle from two lanes to one, so the traffic moves on one lane. In such cases, the maximum capacity of 68% was assumed. For traffic volume below 1000 veh/h time loss is insignificant: 0-12 veh/h in total for all drivers. If the traffic volume is greater than 1000 veh/h, the time loss values grow rapidly. This applies to a traffic volume with capacity as high as 34%.

A traffic incident in which two lanes are blocked occurs when it is not possible to remove vehicles from the roads to allow movement on one lane or move the vehicles to the emergency lane or shoulder. Also, before removing the vehicles from the road, two lanes are blocked. The model assumes that capacity is reduced by 100%. Even for a volume of about 500 veh/h, the blockage on the entire cross-section of the road causes a considerable loss of time. The greater the traffic volume, the quicker the time loss grows. The queue of vehicles, which is formed as a result of the traffic incident disperses over a very long time. This applies to the traffic volume with a capacity of even 11%.

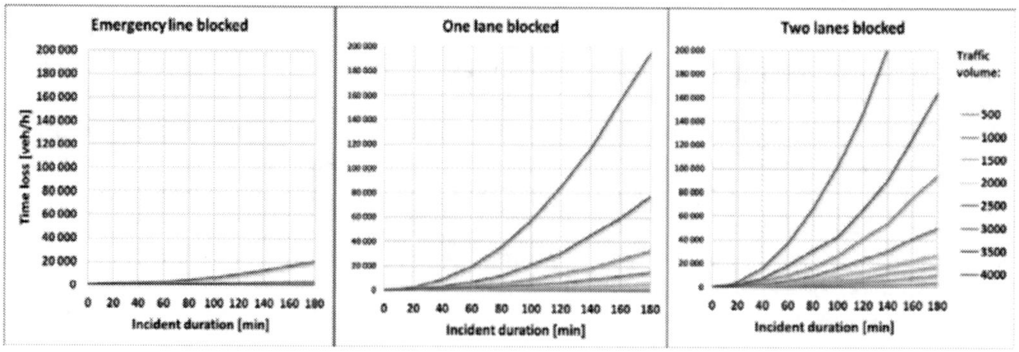

Fig. 3. Time loss graph – reduction variants of 25%, 68% and 100% (own study).

4.3 Time loss per vehicle

Time loss per one vehicle for the variant of a blocked emergency lane is not more than 33 minutes, for the analysed scenarios. If the traffic volume is less than 3000 veh/h the time loss is almost zero, and amounts up to 1 minute. For higher traffic volumes, time loss increases to the point of becoming an impediment for the road users.

If one lane is blocked, the graph differs from the previous variant. For traffic over 1000 veh/h, the loss of time per vehicle begins to grow rapidly. With a further increase in traffic volume, time loss increases to a lesser extent. This is due more to the specifics of the behaviour of the queue of vehicles. The duration of the incident was observed to have a major impact.

If the entire cross-section of the road is blocked, the time loss graph per driver differs from the two preceding variants. In this case, the main factor of time loss is the duration of the traffic incident. It was observed that in every case, the time loss did not exceed the duration of the incident. If the traffic was at capacity, the road users' time loss would amount to the equivalent of the traffic incident duration. The increase in the time loss, per driver, depending on the traffic volume is of very low importance, but the loss is characterised by a modest increase. The graph of time losses for the various variants are shown in Figure 4.

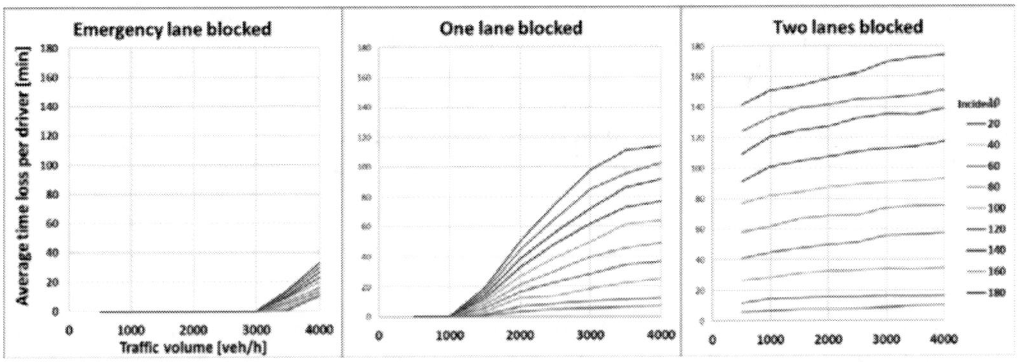

Fig 4. The graphs of average time loss per driver (own study).

5 Time loss model

The analysis and simulation study results made it possible to create a model of time loss depending on the parameters included in the analysis. The model of time loss is a function dependent on the traffic volume on the section, incident duration and level of capacity reduction. In addition to this, there are more factors that affect time loss significantly, which may also be added to the time loss model after further analysis.

The graph of function is variable depending on whether the reduction of capacity results in a so-called "bottleneck", hence causing the phenomenon of choking. If traffic volume is high and the reduction is significant it may cause an increase in the length of the queue which will be almost equal to the traffic volume. Long-lasting traffic incidents with such a reduction generate a significant increase in loss of time. The general function can be shown as:

$$t_s = a \cdot {t_z}^b \cdot e^{(c \cdot S \cdot N)} \tag{1}$$

where:
- t_s - time loss,
- N - volume,
- t_z - incident duration,

- S - level of capacity reduction (relative decrease in capacity),
- C - base section capacity in no incident situation [veh/h],
- a, b, c – coefficients.

6 Conclusions

6.1 Summary

Based on the analysis, it was found that it is possible to map the phenomenon of road users' time loss caused by a traffic incident, using a mathematical model.

It was noted that the most important parameters affecting the total time loss was traffic volume and the level of capacity reduction. These parameters are closely related and have a similar effect on the upward trend in time loss with increasing duration of the traffic incident. A different pattern of time loss increase is related to the extent of the restriction, the nature of the queue and the time needed to restore the traffic conditions to the state before the incident. Due to these differences, the time loss in the model is calculated using three formulas for different levels of restriction.

Loss of time per driver depends on the level of capacity reduction, traffic volume and duration of the traffic incident. The most important factor is capacity reduction because it alters the behaviour of the queue of vehicles. With the increase in restrictions, average time loss rises constantly, though to a decreasing extent. The maximum time loss per driver in given conditions will not be greater than the duration of the incident.

Due to the high importance of the level of capacity reduction for time loss, it is important to remove vehicles from the road or move vehicles to one lane as soon as possible to make at least one lane passable.

6.2 Future research

The proposed model of time loss is rather simplified in spite of the many variables. It refers to a situation where a driver cannot change the route. The analysed section was also simplified to one road with unchangeable cross-section and parameters.

Future research is expected to develop a model with more parameters which will allow a mapping of each traffic incident and estimating the time loss for a particular case. In addition to increasing the number of analysis parameters, it will provide a calculation of the additional variables dependent on the calculated time loss and resulting difficulties, such as increased fuel consumption or increased environmental pollution.

The analysis based on a simulation study is expected to be extended by adding field research and statistical analysis of the road incident database with greater detail than the study presented in the paper. The extended model will also be confronted and calibrated with actual data from national roads.

References

1. R. Kucharski, G. Gentile, Modelowanie zjawiska zmiany trasy przejazdu w dynamicznym rozkładzie ruchu w sieci drogowej, in: Res. Tech. J. Polish Assoc. Transp. Eng. Cracow, (Cracow, Poland, 2014)
2. R. Kucharski, Rerouting phenomena modelling for unexpected events in Dynamic Traffic Assignment. Modelowanie zjawisk zmiany trasy przejazdu w sytuacjach nietypowych, Cracow University of Technology, 2015

3. PB Farradyne, *Traffic incident management handbook* (2000)
4. E. Chin, O. Franzese, D. Greene, H. Hwang, R. Gibson, Temporary losses of highway capacity and impacts on performance (2002)
5. S. Gaca, W. Suchorzewski, M. Tracz, *Road traffic engineering. Theory and practice. Inżynieria ruchu drogowego. Teoria i praktyka*. (WKŁ, Warsaw, 2008)
6. The database of the Pomeranian Voivodship Fire Service, (2016)

Functionality of road safety devices – identification and analysis of factors

Łukasz Jeliński[1,*], *Kazimierz* Jamroz[1], *Jan* Jamroz[1], *and Marcin* Antoniuk [1]

[1]Gdansk University of Technology, Faculty of Civil and Environmental Engineering, Gdansk, Poland

Abstract. Road safety devices are designed to protect road users from the risk of injury or death. The principal type of restraint is the safety barrier. Deployed on sites with the highest risk of run-off-road accidents, safety barriers are mostly found on bridges, flyovers, central reservations, and on road edges which have fixed obstacles next to them. If properly designed and installed, safety barriers just as other road safety devices, should meet a number of functional features. This report analyses factors which may deteriorate functionality, ways to prevent this from happening and the thresholds for loss of road safety device functionality.

1 Introduction

Run-off-road (ROR) crashes, is still of the unresolved road safety issues. The consequences of such event are secondary accidents ending with: turning the vehicle or hitting the vehicle at a facility near the road (tree, pole, ramp). ROR account for more than one third of all road users killed in road accidents in Poland. Road safety devices are a group of devices that reduce the probability and the effects of a vehicle ROR. Road safety devices in this article are: traffic safety equipment (road safety barriers, energy-saving equipment, crash cushions) and traffic layout devices (vertical marking, horizontal marking). These measures must meet the criteria of vitality and many functional properties such as, ,-safety, reliability, efficiency, effectiveness, preparedness, day and night visibility, etc.

2 Defining the functionality of road safety devices

The basic term that helps assess whether a device is fit for purpose is functionality. The functionality (practicality) of road safety devices means that they have to stay in operation 24 hours a day and 365 days a year. Continuity of operation ensures that road users are provided on an on-going basis with the right quality of parameters for communicating and receiving information, for driving a vehicle and with safety.

The functionality of road safety devices is managed by:

1. In the case of road traffic safety devices – maintaining a high level of preparedness (reliability); ensuring that the functionality stays strong if drivers behave the wrong way or vehicles break down (enforced driver behaviour, making sure the vehicle stays on the road, protection of road users' health and life, protection of objects and people on the road); monitoring the basic parameters of the devices (e.g. poor condition, loss of

* Corresponding author: lukjelin@pg.gda.pl

technical parameters) through maintenance, exchange or use of more modern solutions, swift repairs of damage and breakdown.

2. In the case of traffic layout devices – ensuring a well-functioning traffic by making traffic layout devices clearly visible in day-time and night-time; monitoring the basic parameters (e.g. retro-reflective features) through treatment and exchange or application of more modern solutions, swift repairs.

The functionality of safety devices depends on the type and durability of the material they are made of [2], and the conditions in which the devices are deployed (weather, winter maintenance, occurrence of micro crashes by vehicles).

3 Measures and requirements for safety barrier functionality

The functionality of safety barriers is assessed primarily against the standard PN-EN 1317. It is an indispensable document used for developing safe road restraints. It sets out the criteria and methods for crash tests of road restraints, including safety barriers, before they can be used on public roads. While the standard does not specify the size, shape or material of safety barriers, it describes the operating classes of safety barriers by identifying their functionality features such as: containment level, deflection expressed with working width and crash intensity.

Eight most frequent functionality features are assigned functionality measures divided by road safety devices and road sections that have these devices.

Table 1. Total time from the first look until the target is passed and Time during which driver's gaze is fixed on target (ms) in 4 analysed situations [3].

Approach	Individual	Societal
Features of functionality:	Road safety device	Road section
Safety	ASI, THIV, PHD	societal and economic consequences
Reliability	Wm,,VI, Dm	closure time, time lost
Capacity	-	capacity
Maintain and redirect the vehicle	wheel position, displacement, load, course of driving	closure time, time lost
Performance	degree, size of damage	congestion (traffic jams)
Capacity to restore	restorability	closure time, time lost
Economic effectiveness	costs of construction, exchange, removal	costs of accident and time lost
Environmental impacts	absorption of zinc in the ground	emissions, noise levels

The resulting impact accelerations and decelerations are measured using: the Acceleration Severity Index (ASI), characterizing the intensity of the impact, and is regarded as the most important rate of impact on occupants; the theoretical head impact velocity (THIV), describes the theoretical speed of the head, colliding with an obstacle during an impact. It has to be less than 33km/h; and the post-impact head deceleration (PHD) describes the head deceleration after an impact and has to be less than 20g (acceleration of gravity).

The dynamic deflection (Dm), the working width (Wm) and the vehicle intrusion (VI) allow a determination of the conditions for installation of each safety barrier and also to define the distances to be provided in front of obstacles.

4 Methods for assessing the functionality of road safety devices

4.1 Studies of road safety device functionality

US and Swedish research [3–7] shows the importance of repairing damaged safety barriers. Failure to do so may lead to loss of functionality. It is clear that a completely damaged safety barrier must be repaired. Safety barriers are frequently affected by minor damage such as shallow dents, the result of collisions at low speeds and small approach angles. Minor damages may also be caused by routine maintenance work, snow ploughing, mowing operations or variable weather. Regardless of the cause of damage, maintenance services must be able to identify those defects, which if untreated, may have fatal consequences. The authorities responsible for the road may be held liable for failure to keep the safety barriers operational. Clearly, there is a need for guidelines to help with assessing damage to safety barriers. It is important to ensure that road maintenance services have the capacity to identify what may seem as minor damage to road safety devices.

The research of recent years [5,6] was primarily focused on identifying the criteria for assessing the degree of damage to steel safety barriers. A tool was to be developed to show the urgency of putting in a new safety barrier in place of a damaged one. The research methodology was based on a review of the literature and a survey of maintenance agencies to establish the methods they used for safety barrier repairs. The review of the literature looked primarily at available national guidelines for repairing safety barriers and guidelines followed by state road authorities. The guidelines of road agencies would usually have two components: (1) maintenance manuals that described the conditions and requirements for repairs, especially safety barrier repairs and (2) maintenance assessment criteria which are used to assess barrier functionality.

With the literature review as the basis, the NCHRP agency developed and sent 22 questions to road maintenance agencies. The questions were divided into five sections: (1) inventory of guardrail and median barriers, (2) repair policies, (3) non-crash related damage or deterioration, (4) notifications and repair responsibilities, (5) inspection policies and procedures.

The purpose of the barrier inventory section was to understand the types of barriers most used by specific road agencies. The repair policies section was intended to provide insight into what thresholds are currently used to determine barrier repair needs, how damaged sites are prioritized, timelines for repairs, documented cases of impacts into damaged barriers, and whether the agency would benefit from more quantitative barrier repair guidelines. The non-crash section looked at the occurrence of corrosion of steel elements, rotting of wooden elements and loss of tension in the case of lines. The notifications and repair responsibilities section was added to diagnose the procedures applied when repairs are diagnosed formally or informally and who is responsible for the repair. The final point in the survey covered inspection policies and procedures and was designed to collect information about types of inspections and how they are conducted and to understand what maintenance assessment methods are used.

To identify the criteria for assessing the degree of damage to steel safety barriers, crash tests and computer simulations were conducted using damaged barriers. The damages were to reflect a variety of damages normally encountered on the roads. The crash tests were conducted using a pendulum and car. The computer simulations were based on the finite element method in the programme LS-DYNA.

One of the tests analysed the effects of a missing or damaged post on barrier performance upon a crash. To that end a simulation was conducted using the finite element method in the programme LS-DYNA. The simulation was validated based on available literature data from previous crash tests. The simulation was designed to test how many posts can be removed while still keeping the operational performance of the barrier. The simulations were conducted for 1, 2 and 3 missing posts (Fig. 1) with the crashes conducted at two points: at the beginning of an unsupported span and mid-span where the missing post should be.

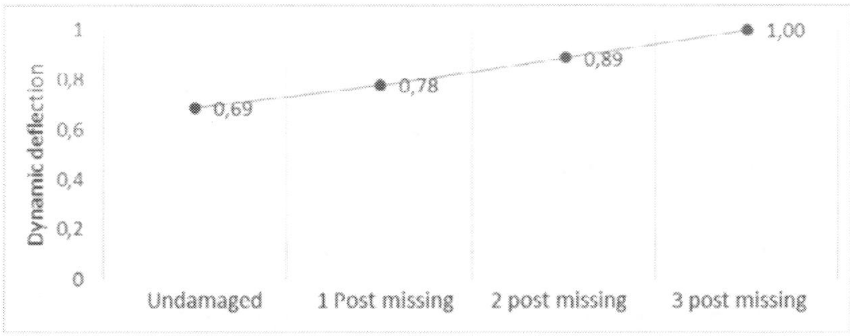

Fig. 1. Simulation results of the barrier crash where the post links into the guardrail.

Based on the results, the scientists concluded that even if only one post is missing/damaged, barrier functionality deteriorates significantly. While none of the vehicles in the simulation overturned, they exhibited significant instability. As a result, such damage was given a high repair priority.

4.2 Guidelines for assessing the functionality of road safety devices

Introduced in 2008, US guidelines [3] for the repairs of steel safety barriers are the US's mandatory document required by the FHWA (Federal Highway Administration). It is dedicated to road maintenance personnel and gives a comprehensive description of the importance and logistics of steel barrier repairs. The guidelines specify when safety barriers must be repaired, which barriers should be exchanged without any delay and classify barrier damage degrees. Damaged barriers are divided into three categories: (1) non-functional, (2) damaged but still functional, (3) functional with minor damage. Table 3 shows a diagram of the method for assessing functionality and damage classification.

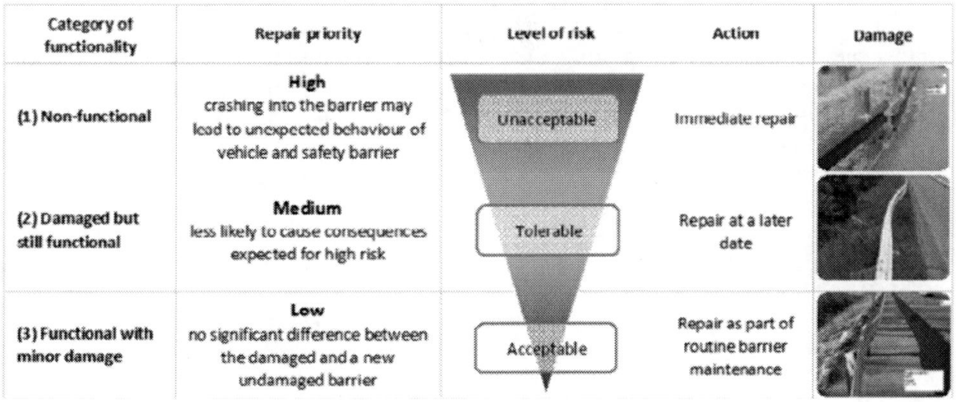

Fig. 3. Method for functionality assessment in the US [3].

US guidelines [3] determine the category of safety barrier functionality based on quantitative data. Category 1 of functionality – barrier is no longer reasonably functional – the rail element is separated (no barrier continuity), partly torn or the rail height is less than 610 mm. In addition, category 1 is when (1) there are three or more broken, bent, or separated posts and the amount out of alignment is less than 305 mm or (2) if the amount out of alignment is greater than 30 cm regardless of post damage. Category 2 of functionality – barrier should function adequately under a majority of impacts – if the guardrail is out of alignment not more than 152 mm and not more than 1-2 posts or when the guardrail is out of alignment not more than 305 mm and not more than 2 damaged posts. Category of functionality 3 – the barrier meets functionality parameters and no replacement is necessary – applies if the posts have no damage and the guardrail is out of alignment by less than 152 mm.

As prescribed by the FHWA guidelines, the type of damage is decisive for whether the device must be repaired or not. The report recommends repairs to Category 1 damage as soon as possible, depending on the hazard it causes for other users. In the case of Category 2 and 3, because the risk is lower, repairs can be made later or as part of routine maintenance.

5 Concept of a functionality assessment method

Today, when maintenance checks are conducted, it is not possible to inspect road safety devices for all the functional features on site. Additional on-site, lab and simulation tests are required. Following from preliminary analyses, a method is proposed for assessing the functionality of road safety devices using a qualitative and quantitative assessment. The qualitative method for assessing road safety device functionality involves a visual check of the condition of the device. The operational performance of a device is established by checking the guardrail for being out of alignment, system integrity, corrosion or damage to protective layers. The quantitative method for functionality assessment should be divided into simplified and detailed. The simplified quantitative method uses the simplest tools for measuring the degree of damage such as depth of deflection, angle of post deflection, etc. The detailed method supports very accurate measurements such as a change in the thickness of anti-corrosion surfaces. There are not many specialist tools available in Poland and worldwide capable of clearly establishing whether a device is functional and to what extent. In this case the quantitative method is justified only if supported with guidelines for assessing the deterioration of road safety devices.

The guidelines should include a classification of damage by type of device and a clear definition of acceptable amounts out of alignment. Equipped with such tools, maintenance services are able to decide on the priority of damaged devices.

Building on the studies and guidelines for assessing safety barrier functionality, work can begin to conduct new research and develop national guidelines for road safety device maintenance. Proposed by the authors, the concept of a method for assessing the functionality of road safety devices involves five stages of tests: 1) Identify the most frequent damage, 2) Assess the functionality of devices using the damage database, 3) Assess the risk involved in each identified hazard, 4) Assess risk acceptability, and 5) Select actions and priorities for repairs.

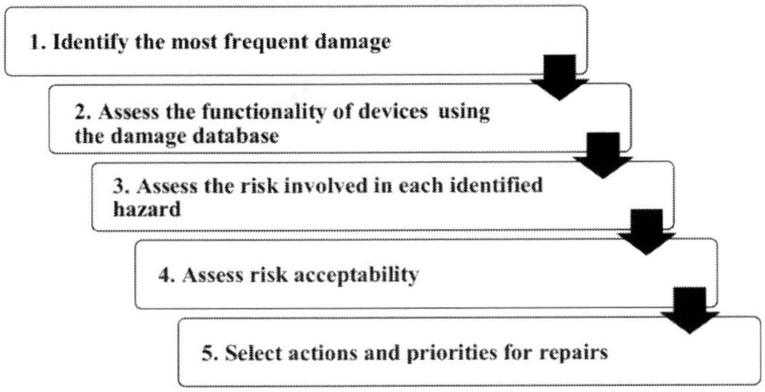

Fig. 2. The concept of a method for assessing the functionality of road safety devices.

Work on the concept of risk assessment is part of a research project for the NCBiR (National Centre for Research and Development) and the GDDKiA (General Directorate for National Roads and Motorways): Project RID 3B "The effects of time and operational conditions on the durability and functionality of road safety elements".

6 Conclusions

Road safety barriers must meet a number of functional features such as: safety, reliability, operational performance, effectiveness, preparedness, etc. Research is needed to tackle unsolved problems, i.e. lack of methods for assessing the effects of safety barrier type, durability and operational conditions on the functionality of the types of safety barriers, lack of methods for selecting the thresholds for assessing the features (functions) of the devices. The most important function of road safety devices is to protect road and roadside users from the risk of injury or death. While there is extensive independent research into the effects of selected factors on road safety device functionality [8], a comprehensive approach is still in short supply.

Acknowledgements

Research conducted under project implemented within RID programme, funded by the Polish National Centre for Research and Development and the General Directorate for National Roads and Motorways in Poland, for the years 2016-2019.

References

1. Journal of Laws of the Republic of Poland, Annex to nru 220, pos. 2181 of 23 December 2003, Detailed technical conditions for road signs, signals, traffic safety equipment and conditions for locating them on the road (2003)
2. NCHRP Report 711, Guidance for the selection, use, and maintenance of cable barrier systems (Washington DC, 2012)
3. FHWA, W-Beam Guardrail Repair: A Guide for Highway and Street Maintenance Personnel (U.S. Department of Transportation, 2008)
4. AASHTO, Maintenance Manual for Roadways and Bridges, 4th Edition (American Association of State Highway and Transportation Officials, Washington, 2007)
5. D. Gabauer, H.C. Gabler, Review of Current Damage-Level Criteria for Longitudinal

Barrier Repair, Transp. Res. Rec. **702** pp. 1–17 (2008)

6. H.C. Gabler, D.J. Gabauer, C.E. Hampton, Criteria for Restoration of Longitudinal Barriers, NCHRP Report 656, t. 22, no. 646 (2010)
7. G. Fredriksson, H.G. Holmen, The ageing of safety barriers . Functional limits for safety barriers that deviate from the nominal in height and / or perpendicular position, (2013)
8. K. Jamroz, M. Budzyński, W. Kustra, L. Michalski, S. Gaca, Tools for road infrastructure safety management - Polish experiences, in: Transp. Res. Procedia, pp. 730–739 (2014). doi:10.1016/j.trpro.2014.10.052

MATEC Web of Conferences **122**, 02006 (2017) DOI: 10.1051/matecconf/201712202006
GAMBIT 2016

Testing the durability and function of road traffic management devices

Łukasz Jeliński[1,*], *Joanna* Wachnicka[1], *Jan* Jamroz[1], *Małgorzata* Kalisz[2], *Piotr* Kaźmierczak[2], *Rafał* Lusa[3], and *Paweł* Skwierczyński[3]

[1]Gdansk University of Technology, Faculty of Civil and Environmental Engineering, Gdansk, Poland
[2]Motor Transport Institute, Warsaw, Poland
[3]Road and Bridge Research Institute, Warsaw, Poland

Abstract. Traffic management measures (vertical signs and horizontal marking, reflective elements) are used for guiding vehicles optically, indicating road mileage, marking objects in road gauge, marking vehicle and pedestrian safeguards and driver information and warning. This paper presents a synthesis of a literature study and the results of research conducted under stage one of the project LifeRoSE. The requirements for different traffic management measurements are described as well as durability and functionality tests and analysis of factors which influence durability and functionality of these measurements.

1 Introduction

The process of installing road safety devices on a road section has several stages. These include planning, design, construction, maintenance and removal of the devices [1]. The lifecycle of a structure involves work which includes planning and design, construction (building) and repairs, maintenance and finally demolition. In addition, these activities can be delivered at different levels of management: strategic (legislation and programming), tactical (planning and design) and operational (construction, repairs and day-to-day maintenance). Decisions that are taken at the strategic and tactical levels are important for the durability, functionality, effectiveness and costs of the devices. Today, those decisions are taken in an environment of scarcity of information, a lot of uncertainty and a lack of the right methods and tools. As a result, the decisions are far from optimal [1].

Durability is defined as the ability of a device or object to maintain its utility over time. In the case of road safety devices used on the road durability means the ability (capacity) to maintain the assumed functionality during the road object's required life. Durability and life expectancy requirements for road markings are defined in standards, guidelines and recommendations which are device-specific [2]. The life expectancy of road safety devices depends on the type and durability of manufacturing material [3], and how the devices are used. While this has been researched at length, it is not clear how the conditions of operation (weather, winter road maintenance, occurrence of micro vehicle impacts) affect the life expectancy of road devices. Road safety devices must meet a number of functional features such as safety, reliability, fitness, effectiveness, readiness, day and night-time visibility, adequacy and others. Some of the undesired problems that require research include a lack of

* Corresponding author: lukjelin@pg.gda.pl

methods for estimating how the type and durability of traffic management devices and operating conditions impact the functionality of the specific groups of road safety devices. Also, there are no methods for selecting the acceptability limits of how to assess the properties (functions) of devices.

The most important function of road safety devices is to protect road and roadside users from death or injury. While there has been extensive independent research into how selected factors affect the properties of road safety devices [4], this has never been a comprehensive effort.

This article describes the results of preliminary studies on the durability of horizontal and vertical marking in relation to normative requirements. The research was done in the laboratory and on test sites. This is the first stage of the research project entitled "Life cost analysis of Road Safety Elements".

2 Testing the durability of horizontal road markings

Horizontal markings can be divided into permanent and temporary markings, depending on their expected application. The technologies available divide the devices further into thin layer marking – with a layer of 0.30 mm to 0.90 mm, thick layer marking – with a layer of 0.90 mm to 5.00 mm, smooth (full), structural and profile marking. Their utility features fall into type I – not visible in wet conditions and during rainfall and type II – visible in wet conditions and during rainfall.
Figure 1 shows examples of thin and thick layer markings and of acoustic lines.

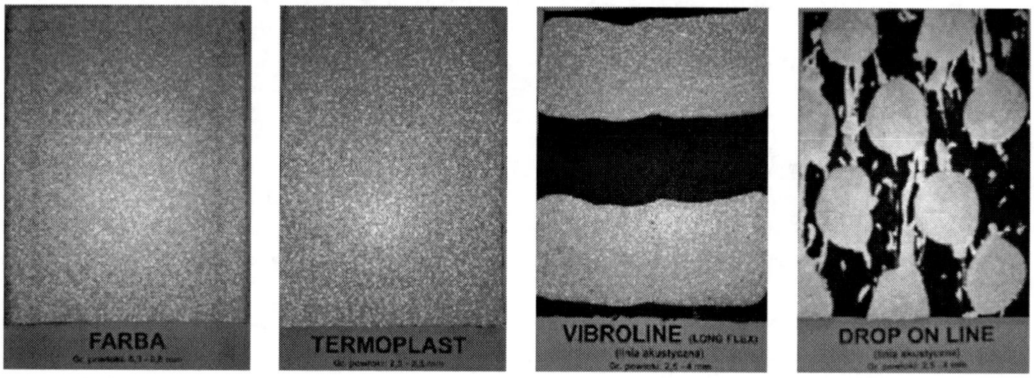

Fig. 1. Examples of horizontal markings.

To assess the durability and function of horizontal marking, the following parameters are used:
- Surface coefficient of retroreflected luminance R_L (in dry and wet condition),
- Coefficient of diffuse luminance Q_D,
- Coefficient of β luminance and chromaticity coordinates x, y,
- Skid resistance tester SRT,
- Class of traffic (number of times a wheel can drive over the marking).

According to the guidelines [5] marking on specific road classes should meet the parameters listed in Table 1.

Table 1. Minimum requirements for permanent horizontal road markings.

Properties	Requirements		
	Motorways	Express roads	Other roads
Luminance coefficient β	0.32	0.32	0.30
Surface coefficient of retroreflected luminance R_L [mcd*m^{-2}*lx^{-1}]	200	150	100
Skid resistance (SRT)	50	50	45
Durability (on LCPC scale)	6	6	6

In Europe two methods for testing durability are used:
- assessment on an experimental section [6] ,
- assessment on a wear and tear simulator [7] Since 1995 Poland has been using experimental sections to test durability: DK 7, DK 5, DK 22, DK 16, and the street Jagiellonska in Warsaw.

Figure 2 shows preliminary results of durability tests of horizontal marking made using thermoplastic materials on the DK 22 experimental section. Lane 1 and lane 2 denote the line number from the edge of the carriageway in the case of the longitudinal durability test pattern [6]. Lane 1 (line next to edge of carriageway) is rarely driven over while lane 2 (line of wheels) is used heavily by traffic. Section 25 is thermoplastic material containing 40% of micro glass beads, section 27 is thermoplastic material containing 30% of micro glass beads. R2 and R3 requirements denote classes of surface coefficient of retroreflected luminance R_L [8] at 100 mcd m^{-2} lx^{-1} and 150 mcd m^{-2} lx^{-}1, respectively.

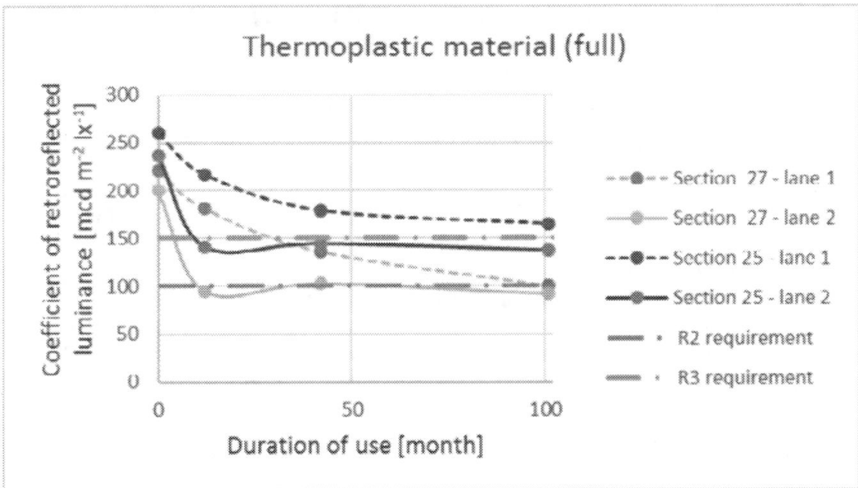

Fig. 2. Test results of retroreflected luminance R_L on DK 22 experimental section depending on how long the marking has been in use.

Figure 3 shows preliminary results of durability tests of horizontal marking made using thermoplastic materials on the DK 16 experimental section. Lane 2 and lane 3 denote the line number from the edge of the carriageway in the case of the longitudinal durability test pattern [6]. Lane 3 (line between wheel tracks) is rarely used while lane 2 (line of wheels) is used heavily by traffic. Section 26 is thermoplastic material containing 30% of micro glass beads, section 27 is thermoplastic material containing 30% of micro glass beads. R2 and R3 requirements denote classes of surface coefficient of retroreflected luminance R_L, at 100 mcd m^{-2} lx^{-1} and 150 mcd m^{-2} lx^{-1} respectively [8].

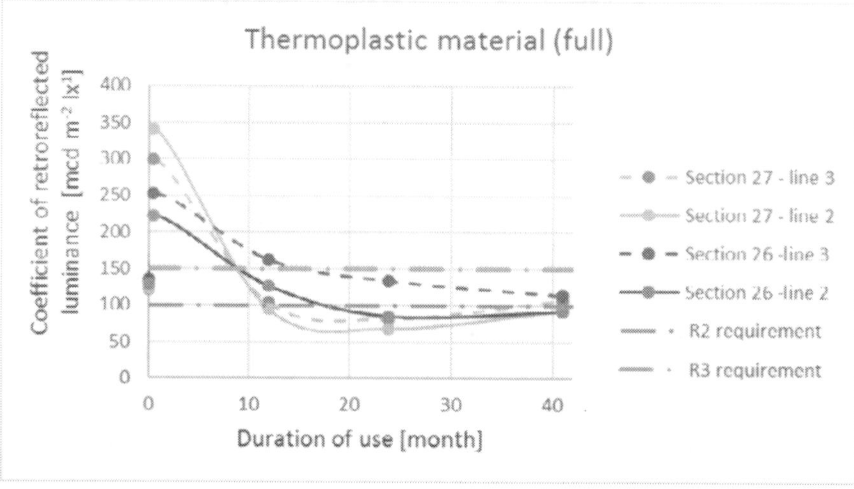

Fig. 3. Test results of retroreflected luminance R_L on DK 16 experimental section depending on how long the marking has been in use.

3 Testing the life expectancy of vertical markings

Permanent vertical road signs are road safety devices in the form of disks and plates with information or symbols, which once installed on a road, have their front surface in upright position. The disk is a flat surface with rigid edges that carry the face of the sign. Disks are usually made of steel zinc coated sheet or powder coated aluminium or otherwise corrosion protected. The face of the sign is the sign's front part designed to inform using symbols or text. It is made of self-adhesive reflective foil, the contents is screen-printed using opaque colour or transparent paint or transparent colour foil. To ensure that the sign is visible from a distance that allows motorists to see the sign, read it and respond appropriately, signs should use reflective materials. The types of reflective materials to be used for specific locations and road class are given in Table 2 [9].

Table 2. Types of reflective foil on road signs depending on road sign location – minimum requirements [9].

Location of sign	**National roads**				**Regional roads**	**County and municipal roads**
	Motorways and express roads	Dual carriageways	International single carriageways	Other carriageways		
Next to carriage-way	2	2	2	1(*)	1(*)	1(*)
Above carriage-way	Prismatic 3	2	2	2	2	2

Reflective foil reflects light in return reflection. Light is reflected directionally and the direction is similar to the incident direction. This property holds even for significantly different incidence directions. The physical parameter which characterises this property is luminance coefficient determined as the relation between the luminous intensity I of a reflective device in the direction of observation and the value of lighting intensity E on that device measured on a plane perpendicular towards the incident light [mcd/lx]. The most frequently used technologies that produce return reflection are micro glass beads (I and II generation foil) and micro prismatic (III and higher generation foil).

Table 3. Type of foil used on road signs in normal and enlarged view.

View	Type of reflective foil		
	I generation foil (type 1)	II generation foil (type 2)	III generation foil (type 3)
Normal			
Enlarged	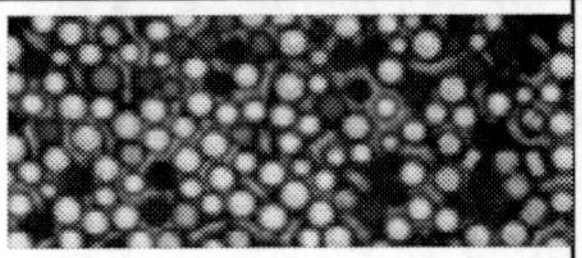	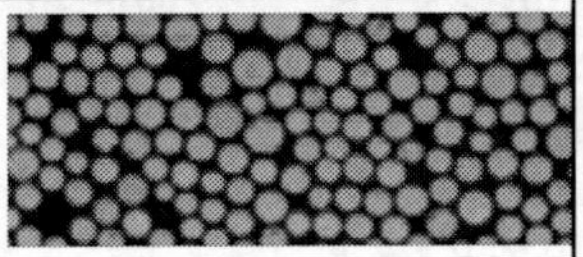	

Between 2005-2010 as part of technical certification, the recommendations of the Road and Bridge Research Institute (the IBDiM) were followed [10]. The Recommendations came as a set of technical and utility properties, research methods and minimal requirements to be met by vertical road signs. The Recommendations take account of the legal regulations [9,11] and are related to material requirements for disks and plates, size and quality requirements, road sign and plate reflective face requirements, operating properties of foil and sign requirements.

Table 4 presents the IBDiM recommendations which also defined the types of materials for disks and plates, types of reflective foil and opaque foil used on sign faces, minimal requirements regarding the surface coefficient of foil retroreflected luminance before and after exposure to water and salt mist (measured in simplified geometry in night-time conditions), tristimulus values requirements (colour) and foil luminance coefficient before and after exposure to salt mist (measured in daylight), operating requirements for foil used on vertical signs and for vertical signs.

Table 4. Normative requirements for reflective foil and vertical signs.

Reflective foil requirements	**Vertical sign requirements**
adhesive power	wind resistance
high temperature resistance	concentrated load resistance
low temperature resistance	resistance to rotational displacement
water resistance	permanent deformation resistance
ball impact resistance	type of face edge
-	drilling through sign face

At present, foils and signs should also be tested for foil retroreflected luminance coefficient in full geometry at night-time, resistance of plates and signs to neutral salt mist and foil durability –accelerated resistance testing of foil to weather conditions in natural and artificial conditions. Manufacturers use different service life durations for reflective foil on sign faces: for type 1 up to 3 years and for type 2 up to 5 years. Please note that in the five years of following the IBDiM Recommendations, the road sign retroreflected luminance coefficient was only tested in simplified geometry, i.e. for a single angle of observation (20') and a single angle of illumination (+5°).

4 Testing the life expectancy of point reflective elements

Road markings also come in the form of point reflective elements installed on the road surface as a complement to horizontal markings. They provide guidance and reflect light to warn, guide and inform road users. They can be built of one or several parts and can be stuck, anchored or inlaid. Reflective points are usually run along the shoulder, on kerbs and lines separating lanes of traffic. Their basic objective is to show how the road runs especially on bends, curves and other sites that require more driver concentration (so called light guidance). Reflective points (permanent and temporary) are tested to ensure comparability and reproducibility of test results. Tests are made on site. The standard [12] gives an exact definition of the requirements to be met by test road sections, weather conditions, traffic volume and type of surface. The standard also defines the minimal time of the test on an experimental road section: at least one year for permanent elements and at least four months for temporary elements. Road tests are conducted in five stages.

1. Assessment of day visibility:
   - check envelope profiles of all elements for sharp edges from traffic side as a result of damage, rubbing or separation,
   - check the elements for integrity, if there are less than 45 elements the test is considered invalid,
2. Assessment of night visibility:
   - check the roadway elements for luminance when lit up with headlights; permanent elements are tested from a distance of 50m ± 3m while temporary elements are tested at 20m ± 2m.
3. Photometric tests, i.e. tests of the retroreflective coefficient and luminance coefficient (for temporary elements only) using elements from experimental sections.

Table 5. Classification of utility properties of point reflective elements in the standard.

Basic assessment:	Night visibility:
Class S0 – unidentified properties	Class R0 – unidentified properties
Class S1- 42 and more of remaining elements	Class R1 – average value of reflection coefficient R 100% of the requirement or more
Class S2 – from 35 to 41 of remaining elements	Class R2 – average value of reflection coefficient R from 50% to 99%
Class S3 – from 1 to 34 of remaining elements	Class R3 – average value of reflection coefficient R from 20% to 49%
-	Class R4 – average value of reflection coefficient R from 1% to 19%
-	Class R0 – unidentified properties

Table 6 lists examples of test results for point reflective elements which the Motor Transport Institute has conducted so far.

Table 6. Motor Transport Institute test results for point reflective elements.

Year of test / service life	Basic assessment	Night visibility ≥ 150mcd/lx
2010 / 18 months	50/50 class S1	8.7 mcd/lx class R4
2011 / 12 months	50/50 class S1	9.0 mcd/lx class R4
2014 / 14 months	47/50 class S1	44.6 mcd/lx class R3
2014 / 14 months, device with ceramic layer	47/50 class S1	260.2 mcd/lx class R1
2014 / 14 months, device in cast iron cover	48/50 class S1	61.6 mcd/lx class R3
2014 / 12 months, device in cast iron cover	47/50 class S1	11.0 mcd/lx class R4

5 Summary

Tests of traffic management devices vary from group to group. The life expectancy and functionality of the devices depends on a variety of factors. The research project LifeRoSE (Life cost analysis of Road Safety Elements) will continue its research on how the function of road traffic devices changes with new samples and test sites to come.

Acknowledgements

Research conducted under a project implemented within the RID programme, funded by the Polish National Centre for Research and Development and the General Directorate for National Roads and Motorways in Poland, for the years 2016-2019.

References

1. K. Jamroz, Method of risk management in highway engineering. Metoda zarządzania ryzykiem w inżynierii drogowej (Gdansk University of Technology, 2011)
2. G.B. Thomas, C. Schlotz, Durable, cost–effecitve pavement markings. Iowa DOT Project TR–454 (Annes, Iowa, USA, 2001)
3. D. Marzougui, U. Mahadevaiah, F. Tahan, C.D. Kan, R. McGinnis, R. Powers, Guidance for selection, use and maintenance of cable barrier system. NCHRP Raport 711 (Transport Research Board, Washington, USA, 2012)
4. K. Jamroz, M. Budzyński, W. Kustra, L. Michalski, S. Gaca, Tools for road infrastructure safety management - Polish experiences, in: Transp. Res. Procedia, pp. 730–739 (2014). doi:10.1016/j.trpro.2014.10.052
5. Z. Szczepaniak, Vademecum poziomego oznakowania dróg. (IBDiM, Warsaw, 2015)
6. BS EN 1824:2011 Road marking materials. Road trials (2011)
7. BS EN 13197:2011+A1:2014 Road marking materials. Wear simulator Turntable (2011)
8. BS EN 1436:2007+A1:2008 Road marking materials. Road marking performance

for road users (2007)

9. Regulation of the Minister of Infrastructure of 3 July 2003 on detailed technical conditions for road signs and signals and traffic safety devices and on the conditions for their location on the road, OJ 2003 No. 220 Pos. 2181 as amended (Poland, 2003)
10. Z. Szczepaniak, Stałe odblaskowe znaki drogowe, urządzenia bezpieczeństwa ruchu drogowego U3, U4, U6, U7, U8, U9, U20, U21, U26, U27, oraz znaki dodatkowe AT, BT, R i W. Zalecenia IBDiM do udzielania Aprobat Technicznych Nr Z/2005-03-009, in: (IBDiM, 2005)
11. PN-EN 12899-1:2010 Fixed vertical road traffic signs. Part 1: Fixed road signs (Poland, 2010)
12. PN-EN 1463-2:2003 Horizontal marking materials - Point reflective elements - Part 2: Field research (Poland, 2003)

Identification and analysis of factors affecting the durability of steel road safety equipment

Joanna Kobus[1,*], *Lech* Kwiatkowski[1], *Rafał* Lutze[1], *Jan* Jamroz[2]

[1]Institute of Precision Mechanics, Warsaw, Poland
[2]Gdansk University of Technology, Faculty of Civil and Environmental Engineering, Gdansk, Poland

Abstract. Factors affecting the durability and life of steel road safety equipment have been reviewed. Based on a literature review and own experience, the following factors have been identified: corrosion, mechanical damage and abrasion during usage and maintenance. Corrosion factors have the biggest impact. Preliminary studies on the annual corrosion losses of selected metals and atmospheric corrosion taking into account regional divisions and roadsides show that the current relations do not apply to immediate surroundings with air pollution emitted by road transport. More research is required on that matter. Local corrosion induced by mechanical damage (gravels, stones impacts) to protective coatings and its impact on steel road safety equipment's durability have also been highlighted.

1 Introduction

The precise description of physical and chemical processes responsible for the phenomenon known as atmospheric corrosion is difficult due to the complex nature of the environment in which the climatic parameters and pollution of the atmosphere change periodically in a chaotic manner with parallel and simultaneous chemical and photochemical reactions [1–3]. The rate of these reactions depends on the type and concentration of pollutants emitted by different anthropogenic or natural sources, the presence of catalysts, change of climatic conditions such as air temperature, humidity, pressure, sunshine, direction and speed of wind. Atmospheric corrosion is usually treated as an electrochemical process, since most of the reactions proceed in aqueous solution and more precisely under a thin film of moisture [4,5].

Understanding the environmental effects of corrosion on materials plays an essential role in technical and economic decisions relating to serviceability of steel road safety equipment. The durability and life of road barriers, poles, sign posts, road fencing or anti-glare screens and other structures is used to qualify the life time of road safety equipment including recommendations for maintenance or even cost analysis.

The data needed for the evaluation, forecasting and empirical description of atmospheric corrosion comes from corrosion monitoring which is defined as a method of constant observation, description and/or measure of the progress of corrosion. A form of standardized monitoring of atmospheric corrosion relies on the use of coupons made from steel, zinc, copper and aluminium exposed in selected areas. In connection with weather and air pollution data collected during annual or long-term exposure the corrosion loss of coupons is determined according to PN-EN ISO standards [6–10]. This helps to classify the corrosivity

* Corresponding author: asia@imp.edu.pl

DOI: 10.1051/matecconf/201712202007

of the atmosphere in accordance with PN-EN ISO 9223:2012 standard. There are five main categories of corrosivity connected with corrosion losses of metals exposed to selected environments and ranging from C1 (very low) to C5 (very high) with low, intermediate and high grades in between. These numbers serve as a recommendation of thickness for zinc coating for galvanized steel or type and thickness of a protective system for paint coatings. They also provide a valuable set of reference data for use in modelling of corrosion rates using dose-response equations. Equations given in PN-EN ISO 9223:2012 have been verified and adapted to the area of Poland during research done at the Institute of Precision Mechanics. An application of the spatial visualization software helped to assess a spatial distribution of atmospheric pollution concentration and corrosion losses across the country. Examples are given at [www.ck]. As a result it was concluded, that atmospheric corrosion of steel, zinc and galvanized steel in the county division did not exceed C3 category, which means medium corrosivity and in many regions was even lower (C2).

Taking into account the level of corrosion damage to infrastructure and new structures, especially steel road safety equipment, research on atmospheric corrosion in direct vicinity of streets, street canyons and highways has been undertaken. The aim of this paper is to present preliminary results, that are the starting point for a complex evaluation of operating conditions and an opportunity for optimising the technical and economic management of road safety equipment.

2 Air pollution in the area surrounding streets and roads

A special case of the environment of a complex chemical nature is the area of big cities where concentrations of pollutants emitted by different sources and temperature conditions are far more diverse and variable than in areas outside the city. This is also the case with roads with heavy traffic. In these areas the exhaust gases from transport are considered the main pollutant. Table 1 lists the content of chemical compounds and particular groups of compounds in total emission from vehicles.

Table 1. The contribution of inorganic and organic compounds released from transport, in 2013 [11].

Lp.	Type of contamination	Percentage in total emission,%
1.	SO_2	0.14
2.	NO_2	32.0
3.	CO	20.2
4.	NMLZO (non-methane volatile organic compounds)	22.0
5.	Particulate matter	19.0
6.	PM_{10}	7.7
7.	$PM_{2,5}$	11.0
8.	hexachlorobenzene	15.0
9.	Polychlorinated biphenols	8.0
10.	Polycyclic aromatic hydrocarbons	1.6

Substances originated from exhaust fumes, as well as those created during vehicle movement by tires and brakes are presented in Table 1. Among them the biggest contribution is from: NOx, CO and solid particles with average grain size of 10 µm and 2,5 µm, marked as PM10 and PM2,5, soot and organic compounds (aliphatic carbohydrates, formaldehyde, aromatic carbohydrates single- and polycyclic and their derivatives). Most of the solid particles are produced by tires and brakes. They are composed of aromatic carbohydrates and heavy metals such as Pb, Cd, Ni and Cu.

The highest concentration occurs in the case of nitrogen oxides. Reactions that take place in high temperature between oxide and nitrogen originated from the air and fuel compounds lead to nitrogen oxide (NO) synthesis. In the next stage NO is oxidized to nitrogen dioxide and other oxides with the participation of UV light and carbohydrates. Nitrogen oxide and dioxide concentration in fumes are thermodynamically and kinetically limited by reagent concentration, flame temperature and temperature in individual engine compartment combustion zones. Average annual NO2 concentration in different areas of Poland is presented in the diagram in Figure 1. During the last 15 years it has stood at a constant, characteristic level, being the lowest in rural areas, about 6 times higher in urban areas without transport. In the area located directly near streets and roads nitrogen oxides concentration is about 2-3 times higher than in urban areas without transport (Figure 2).

The research results including the analysis of nitrogen oxides concentration in urban areas show that quantitative composition of NO2, NO, O3 and organic compounds mixture is characteristic for given area during a specified time. A substantial contribution comes from concentration of oxidants and radiation energy as a direct trigger of photochemical reactions.

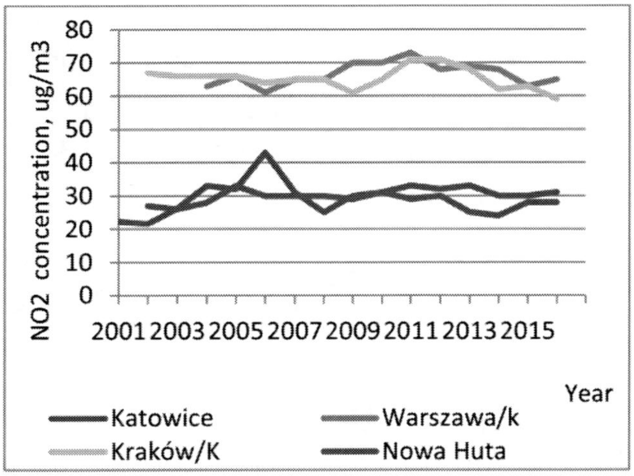

Fig. 1. The annual average concentration of NO2 in the years 2001-2016 for selected urban areas.

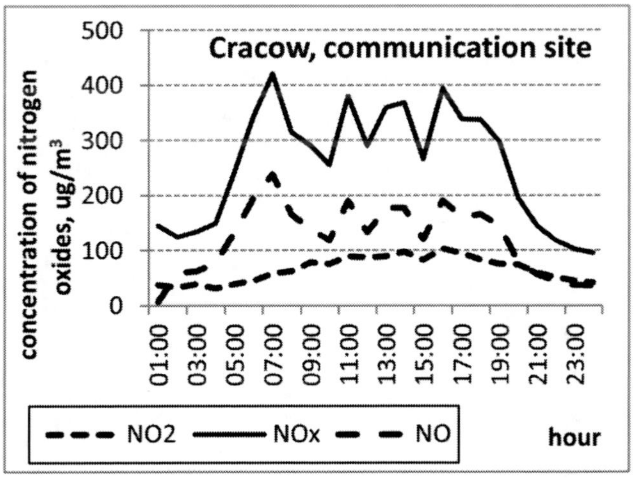

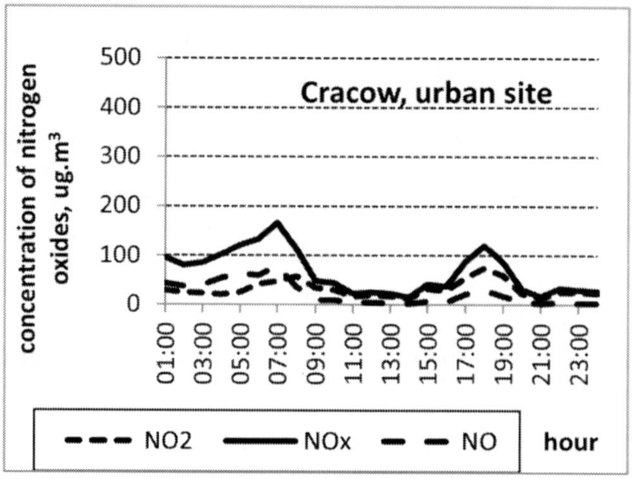

Fig. 2. Changes of daily concentrations of NO, NO2 i NOx for urban and communication corrosion sites in Cracow (25 June 2012 r.) (www.krakow.pios.gov.pl.)

A distribution of pollutants near roads and highways proves that the concentration of gases and solid particles significantly decreases with distance from the centre of a road (Figure 3 A). It also depends on the height of the emission source. Emission size at a given point or on a road section depends on vehicle velocity, engine and fuel type. As can be seen in Figure 3B the same trend is observed for both nitrogen concentration and corrosion loss of zinc.

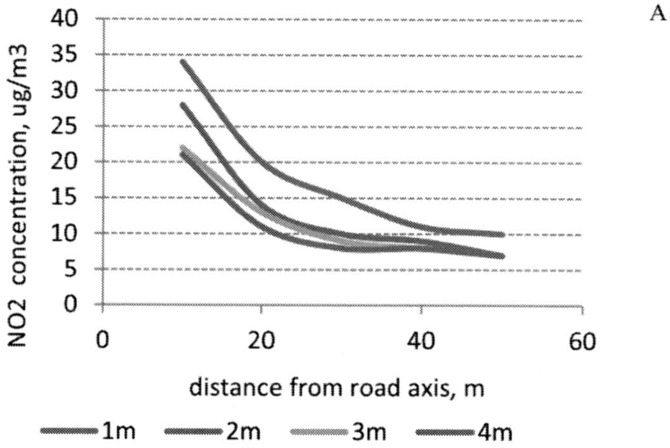

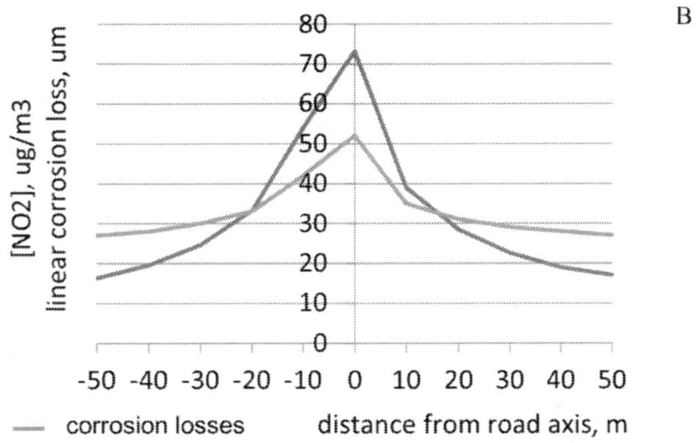

Fig. 3. A) a change of NO2 concentration with distance from the road axis and height of emitter (data related to vehicles driving at 110 km/h), B) a change of NO2 concentration with distance the road axis and related change of corrosion loss of zinc coating.

3 Preliminary results and discussion.

Test specimens of dimensions 150 x 100 x 1 mm were made of the following materials:
- low carbon steel sheet according to PN EN 10139:2016-04, grade DC05
- zinc sheet according to PN EN 988 / PN EN 1179:2005, min. zinc concentration 97.7%
- hot dip galvanized low carbon steel sheet coated in batch process.

Experimental racks were assembled at a distance of 0.5 m to 4 m from the edge of the carriageway next to streets with different traffic levels (115/5000 vehicles per hour). Corrosion losses were determined after annual exposure of the samples according to PN-EN ISO 9224 and atmospheric corrosivity categories were established according to PN-EN ISO 9223: 2012. The results are presented in Table 2.

Table 2. Changes of atmosphere corrosivity at selected sites in the vicinity of roads for steel and zinc determined in annual exposures in the years 2014 – 2016.

Exposure site	Zinc			Steel		
	2014	2015	2016	2014	2015	2016
Niepodległości Av.	C3	**C5**	-	C2	C2	C3
Gen. Maczek St.	C3	C3	**C5**	-	C2	C2
Skłodowska-Curie Bridge	**C4**	C3	**C5**	C3	C3	C3
Czerniakowska St.	C4	C3	**C5**	C3	C3	C3
Siekierkowski Bridge	C5	C3	**C4**	C3	C3	C3
Viaduct at Central Station	C4	C4	**C5**	C2	C3	C3
Primate of Millenium Av.	C4	C3	**C5**	C3	C3	C3
Toruńska Route	-	**>C5!**	**>C5**	C3	C4	C3
Tunnel at Wisłostrada	C4	**>C5!**	**>C5**	C2	C4	**C4**

Wisłostrada/Bielany	C4	C3	**C4**	C3	C4	**C4**
Wisłostrada/Cytadela	-	C3	**C5**	-	C3	C3
Łazienkowska Route	-	**C4**	**C5**	-	C3	C3
Katowice/A4	**C4**	**C4**	C3	C3	**C4**	C3
Katowice/ Urban	C3	C2	C3	C3	C3	C3
Cracow/Nowa Huta	C3	C2	C2	C3	C2	C2
Cracow/ communication route	**C4**	C3	C3	C3	C3	C2

It is notable, that corrosion categories determined from corrosion losses of steel are lower than those of zinc. This is because of a greater susceptibility to corrosion of zinc in the presence of nitrogen oxides and nitric acid in the air. As an example, there are very high zinc corrosion rates in tunnels (tunnel at Wislostrada) and under roofing (Torunska route, viaduct near CS) where the corrosivity category reached C5 and above its upper limit. In areas outside any transport routes (Cracow/Nowa Huta) the corrosion rate of zinc and steel corresponds to C2 and C3 category, which is in compliance with atmospheric corrosivity determined across the whole country.

The question arises about the correlation between causes and effects. The results collected from 58 traffic corrosion sites are expressed by Pearson's correlation coefficients and presented in Table 3. The correlation between the average annual NO2 concentration measured at test sites and the average daily traffic volume for vehicles for two different time periods remain at moderate levels.

Table 3. Pearson's correlation coefficients (r) between traffic intensity and NO2 concentration, as well as corrosion losses of zinc or steel.

No. of sites/ Time of exposure		$[NO_2]$, µg/m^3	Corrosion losses Zn, µm	Corrosion losses Zn coating, µm	Corrosion losses of steel, µm
1 (+N=26) 2 years	Traffic intensity vehicles/h	0,58	0,59	0,34	0,17
2 (+N=58) 3 years	Traffic intensity vehicles/h	0,63	0,26	0,42	0,17

Correlation coefficients between traffic and corrosion of zinc or zinc coating exhibit moderate linear dependence for the time period of 2 years and poor for 3 years, indicating that the corrosion rate can be influenced by local factors not taken into account in the analysis, e.g. wind speed and direction, type of vehicles and fuel, vehicle speed etc. Another reason may be an insufficient population of test sites and too short period of observations.

As it was stated at the beginning of this paper the spread of pollutants is mainly influenced by many agents and parameters which in turn affect corrosion rates of materials. The most important are type of pollutant and volume of emission, type and the origin of emitter and its parameters, shape and topography of the area characterized by roughness of terrain, meteorological conditions including circulation type of atmospheric content, atmospheric equilibrium, wind direction and velocity, air temperature and its vertical gradient, size and type of precipitation near the emission source and near the receptor, intensity of sunlight, cloudiness and humidity.

To find a solution to the possible relations between local road factors listed above and corrosion losses of structural materials would require extremely intensive research which would be very difficult to complete in a reasonable period of time. At least for financial and logistic reasons. Taking into account a phased approach, the most urgent studies should focus on a more precise explanation followed by relevant equations of the relation between corrosion rate of road safety elements and the intensity of traffic. This will lead to the development of a software as a tool for the adequate prediction of corrosion behaviour of infrastructure at local sites. Consequently, it will help to formulate practical recommendations to extend the durability and life of road safety elements.

4 Conclusions

1. A comprehensive review of scientific and technical literature supported by the authors' own research revealed that pollution emitted from heavy traffic affects to a greater extent corrosion of metals than many other industrial corrosion agents. The origin of chemical compounds comes from exhaust gases, tires and brakes. In order to achieve an optimum durability of road infrastructure a determination of corrosivity of the atmosphere in the vicinity of highways and roads seems to be the most serious and urgent problem to solve. This kind of research has not been carried out in Poland.
2. It is proved that dose – response equations for modelling the atmospheric corrosion rate developed for relatively large administrative areas and proposed in PN EN ISO standards do not apply to the local sites where the number of parameters can be greater and the dynamics of events is much more intense in comparison with regional area. This conclusion shall be assumed as the genesis of this work devoted to the determination of corrosivity of atmosphere in the vicinity of roads and highways with special attention paid to the characteristics of sources of corrosion agents and traffic intensity in particular.
3. The corrosivity of local atmosphere along roads is more aggressive towards zinc than steel. Zinc is the major protective material against corrosion, usually applied as zinc coating on steel. It is confirmed that corrosion rate of zinc is 3 – 5 times faster within a road area than in a regional division which poses additional requirements for surface finishing of structures exposed to local corrosion.
4. Corrosion category in big cities is the highest for zinc and zinc coatings located in sheltered and/or screened sites, e.g. in tunnels but one may find similar areas across the country. Therefore a new exposure started in 2016 as part of this work. It is located in environments containing sheltered and un-sheltered conditions with a diversified corrosion load.

Acknowledgements

This work was supported in part by the National Centre for Research and Development and General Director for National Roads and Motorways – research programme „The Development of Road Innovations, project RID 3B (agreement nr TZP/RID-I-64/12/NCBR/2016) NCBiR and GDDKiA in the frame of LifeRose Project.

References

1. R. Westerholm, K.E. Egebäck, Exhaust emissions from light- and heavy-duty vehicles: chemical composition, impact of exhaust after treatment, and fuel parameters, Environ. Heal. Perspect. **102** pp. 13–23 (1994)
2. A. Venkatram, V. Isakov, J. Yuan, D. Pankratz, Modeling Dispersion at Distances of Meters from Urban Sources, Atmos. Environ. **38** pp. 4633–4644 (2004)
3. K.H. Kwak, B. J.J., L. K.Y, Dispersion and photochemical evolution of reactive pollutants in street canyons, Atmos. Environ. **70** pp. 98–107 (2013)

4. F.D. Wall, M.A. Martinez, W.A. Missert, R.G. Copeland, A.. Kilgo, Characterizing corrosion behavior under atmospheric conditions using electrochemical techniques, Corros. Sci. **1** pp. 17–32 (2005)
5. H. Kaesche, Corrosion of Metals. Physicochemical Principles and Current Problems (Springer, 2003)
6. ISO 9223:2012, Corrosion of metals and alloys - Corrosivity of atmospheres - Classification, determination and estimation (2012)
7. ISO 9224:2012, Corrosion of metals and alloys - Corrosivity of atmospheres-Guiding values for the corrosivity categories (2012)
8. ISO 9225:2012, Corrosion of metals and alloys - Corrosivity of atmospheres-Measurement of environmental parameters affecting corrosivity of atmospheres (2012)
9. ISO 9226:2012, Corrosion of metals and alloys - Corrosivity of atmospheres -- Determination of corrosion rate of standard specimens for the evaluation of corrosivity (2012)
10. PN-EN ISO 8565:2011, Metals and alloys — Atmospheric corrosion testing — General requirements (2011)
11. Krajowy Ośrodek Bilansowania i Zarządzania Emisjami (KOBiZE), Krajowy bilans emisji SO2, NOx, CO, NH3, NMLZO, pyłów, metali ciężkich i TZO w układzie klasyfikacji SNAP I NFR Title (Instytut Ochrony Środowiska – Państwowy Instytut Badawczy, 2015)

MATEC Web of Conferences **122**, 02008 (2017) DOI: 10.1051/matecconf/201712202008
GAMBIT 2016

Tools for road infrastructure safety management in Poland

Wojciech Kustra[1,*], and *Lech* Michalski[1]

[1]Gdansk University of Technology, Faculty of Civil and Environmental Engineering, Gdansk, Poland

Abstract. Road safety can be improved by implementing principles of road safety infrastructure management (RIS) on the network of European roads as adopted in the Directive. The document recommends that member states should use tried and tested tools for road safety management such as: road safety impact assessment (RIA), road safety audit (RSA), safety management on existing road networks including road safety ranking (RSM) and road safety inspection (RSI). The objective of the methods is to help road authorities to take rational decisions in the area of road safety and road infrastructure safety and understand the consequences occurring in the particular phases of road life cycle. To help with assessing the impact of a road project on the safety of related roads, a method was developed for long-term forecasts of accidents and accident cost estimation as well as a risk classification to identify risks that are not acceptable risks. With regard to road safety audits and road safety inspection, a set of principles was developed to identify risks and the basic classification of mistakes and omissions.

1 Introduction

Poland continues to be one of the European Union's worst performing countries for road deaths. In 2015 there were 2,938 people killed on Polish roads with 39,800 people injured. While the priorities set out in national and regional road safety programmes [1] help to systematically reduce Poland's road deaths, the results are far from what is expected. Road safety can be improved by implementing principles of road safety infrastructure management (RIS) on the network of European roads as adopted in the Directive [2]. RIS management involves the use of procedures throughout the life cycle of a road. The purpose of the procedures is to identify road hazards systematically, assess the possible consequences for road users, use measures to eliminate the hazards or mitigate the consequences. The consequences are measured with the number of accidents, injured and killed in road accidents and the costs of road accidents. The document recommends that member states should use tried and tested tools for road safety management such as:
- road safety impact assessment (RIA),
- road safety audit (RSA),
- safety management on existing road networks:
  - road safety ranking (RSM),
  - road safety inspection (RSI).

* Corresponding author: castrol@pg.gda.pl

To meet the needs of road authorities (national, regional and local), the Gdansk University of Technology (Department of Highway Engineering) in cooperation with the Krakow University of Technology (Department of Road Construction and Road Traffic Engineering) have developed several basic tools for managing the safety of Poland's road infrastructure [3–5]. Tools for managing safety on road networks have been developed by a team at the University of Science and Technology in Bydgoszcz [6].

2 Methodology basis

The purpose of the method is to help road authorities to take rational decisions about road safety, road infrastructure safety and the consequences that occur in the various stages of the life cycle of a road structure [1]. Work on building the particular elements of Poland's road infrastructure safety management adopted the following assumptions:
- the management system and its elements will cover all stages of a road structure's life cycle (planning, design, construction, operation and closure),
- road safety infrastructure management is based on risk management,
- a variety of methods to identify hazards and sources of hazards will be used.

A risk management method is a repetitive procedure designed to reduce road traffic risk effectively and efficiently with particular emphasis on interventions and measures related to road infrastructure within reasonable limits. The proposed risk management method includes two distinctive phases:
- risk assessment phase, process of analysing and determining acceptable risk taking into account standards for risk acceptance,
- risk response phase with three important phases: handling risk, monitoring risk and communicating risk.

The practice of road safety improvement just as in medicine, uses two types of therapy: treating the symptoms or treating the causes.
- Treating the symptoms means to identify the risk of hazards and their sources by analysing past road accidents and their causes.
- Treating the causes means to identify hazards and sources of hazards by conducting inspections in the field (on an existing road) or auditing designs (a planned road or one being designed) and by analysing expected (forecasted) road accidents and their consequences.

3 Method of impact assessment

Assessing the impact of a planned road on road safety involves a strategic analysis of how the variants of a road will affect road safety on a network of public roads within the planned road's impact area [7]. The Road Impact Assessment (RIA) is conducted to rank the variants of the planned road by their impact on road safety within the network of roads that are within the planned road's impact zone. The results of the analysis should be included in a multi-criteria analysis (together with other criteria: technical, economic and environmental criteria) when assessing the variants of the road under analysis. The road safety impact assessment should also be used to reject from further design stages those variants that do not meet basic road safety standards. The research problem was to develop a method for forecasting road safety measures such as accident density AD, injury density ID and killed density KD also known as measures of societal risk. Measures of societal risk are calculated using the following relations (example for accident density AD) (1):

$$AD_{i,j,v,k} = \beta_{1,A,1} \cdot Q_{i,j,v}^{\beta_{2,A,k}} \cdot exp\left(\beta_{3,A,k} \cdot Q_{i,j,v} + \beta_{4,A,k} \cdot PHV_{i,j,v}\right) \cdot f_{TP} \cdot f_{RL} \cdot f_{AE} \cdot f_{DI} \quad (1)$$

where:

- $Q_{i,j,v,}$ – average annual daily traffic on the analysed road section j, for the year of the forecast i, variant v (thou. veh./ 24h),
- $PHV_{i,j,v}$ – share of heavy vehicles (trucks and buses) on the analysed road section j in forecast year i, for variant v (%),
- $\beta_{1,} \dots \beta_{n,}$ – equation coefficients,
- k – number of carriageways, k=1 one carriageway, k=2 two carriageways,
- f_{TP} – rate of the effect of the year of forecast which takes account of the level of socio-economic development of a country and systemic actions designed to improve road safety,
- f_{RL} – rate of the effect of road location (curvature, waviness, region of the country),
- f_{AE} – rate of the effect of the type of roadside (urban, industrial, rural, wooded) cut across by the analysed road section,
- f_{DI} – rate of the effect of junction or interchange density DI.

Fig. 1 shows selected road safety measures depending on ADDT intensity. The values for road class A and S are clearly lower.

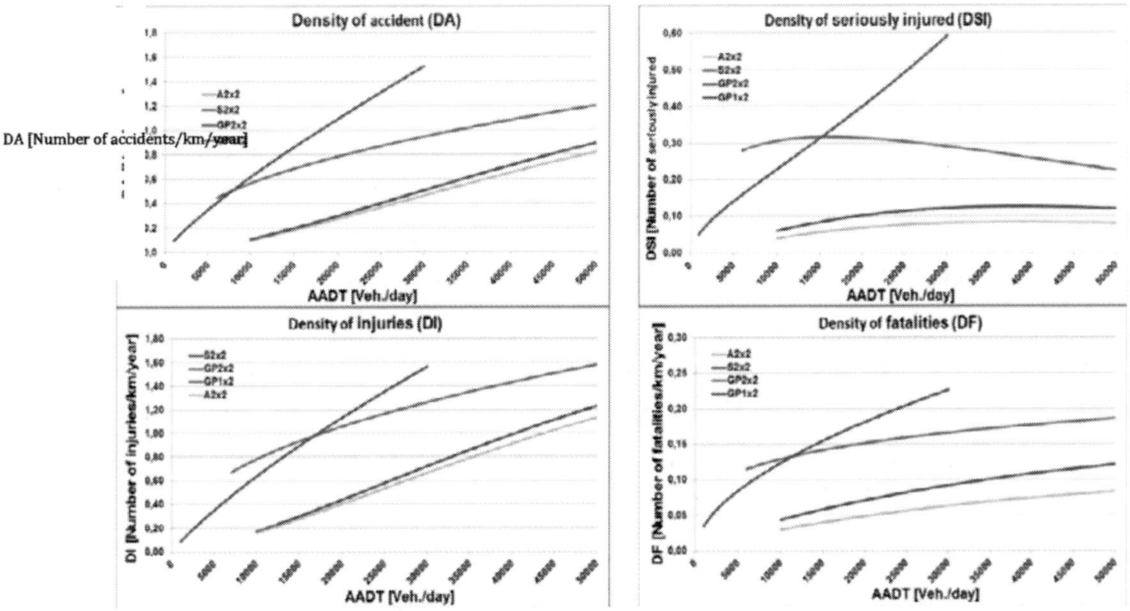

Fig. 1 Road safety measures depending on ADDT intensity.

The authors recommend using the PTV Visum software as one of the elements of the tool Safety PL – Support Tool for Road Safety Impact Assessment. The choice of the PTV Visum software has been dictated by the fact that it is the most commonly used tool for work related to forecasts and analyses of traffic on newly designed roads in Poland. This approach will help to use the results of traffic forecasts for the entire impact area of the planned road, in the prepared models predicting accidents and casualties without the necessity to transfer them to other tools that support the calculation of road safety assessment. The essential element preceding the calculation of road safety measures is preparation of data on homogeneous sections in the impact area. For this purpose, the PTV Visum programme will prepare attributes for all variables in the prepared prediction models of accidents and victims. The modules are shown in Fig. 2.

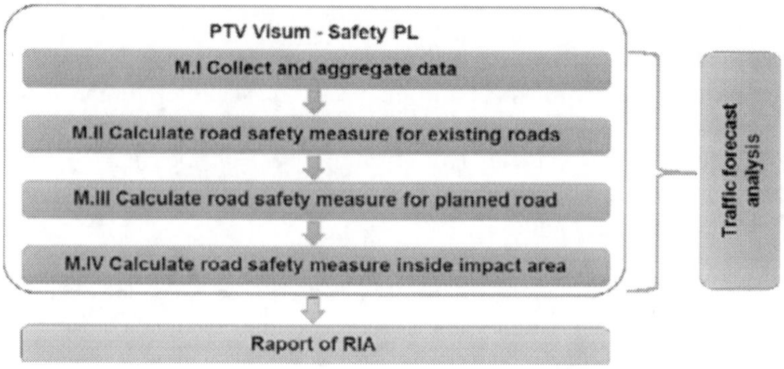

Fig. 2 Safety PL modules.

4 Road safety audit

With more than ten years of audit implementation experience in Poland [7] two groups of problems can be identified. They are related to:
- the process of road design and use of safety standards,
- the prevalence, correctness and effectiveness of the auditing procedures.

In the first group of problems, road safety audit shows that statistically the same errors keep appearing quite often when it comes to designing the cross-section and vertical alignment, layout, junctions and interchanges:
- the use of 1x4 and 1x6 cross-sections with no central reservation,
- structures (utility poles, barriers) are placed on narrow pavements (1.5-2.0 m),
- cyclists and pedestrians are not effectively segregated in the street cross-section,
- steep slopes are used in hazardous places,
- sight distance is not sufficient on horizontal and vertical curves,
- the distances between junctions are too small; junctions are classified in the design as exits
- poor surface drainage of the carriageway,
- selection of the wrong junction type,
- interchanges not matching traffic parameters.

The second group of problems arise due to difficulties with ensuring:
- professional staff and independence of auditors' comments,
- objectivity in assessing a design's proposals for their safety and reasonable recommendations.

At present, there are three documents that are related directly to audit procedures:
- Regulation 42 of the Director General for National Roads and Motorways of 3 September 2009 concerning the road safety impact assessment and road safety audits of road infrastructure projects,
- Act of 13 April 2012 revising the public roads act and some other acts introducing the road safety audit – an independent, detailed and technical assessment of a public road being designed, built, improved or used for the safety of road users,
- Ordinance of the Minister of Transport, Construction and Maritime Economy of 14 September 2012 concerning training and certificates for road safety auditors.

5 Managing the safety of an existing road network

The main objective of road safety ranking (RSM) is to select sections that carry the highest individual risk, i.e. the likelihood of being involved in fatal crash of a road user and sections

that carry the highest societal risk and the biggest potential for reducing accident costs as a result of road authority actions [8]. The intermediate goals of RSM are to:
- systematically assess safety on existing road networks,
- identify and rank high risk sections,
- identify and rank sections with the highest density of accident costs and sections with the highest potential to reduce accident costs,
- create a basis for selecting sections that need work of the highest effectiveness.

In the traffic safety ranking five classes are proposed depending on the potential for reducing accident costs on road sections (A, B, C, D, E) [9]. While hazardous sections must be ranked on national roads only, in 2015 a new ranking was developed for the National Road Safety Council covering regional roads [10]. Fig. 3 shows an example of the ranking looking at societal risk (density of accident costs) for run-off-road accidents.

For the particular technical classes of national roads the risk of an accident was assigned at three levels of acceptance (Table 1):
- unacceptable risk level on a road section means a strong likelihood of severe personal or economic consequences – the road section cannot operate safely until that risk is reduced or the sources of the hazard are removed,
- tolerated risk level on a road section means a medium or low likelihood of personal or economic consequences – the road section may operate temporarily or under certain conditions (such as the use of ad hoc solutions to improve safety such as speed limits, a more intensified road traffic enforcement),
- acceptable risk level means low or very low likelihood of personal or economic consequences – the road section can operate with no additional measures.

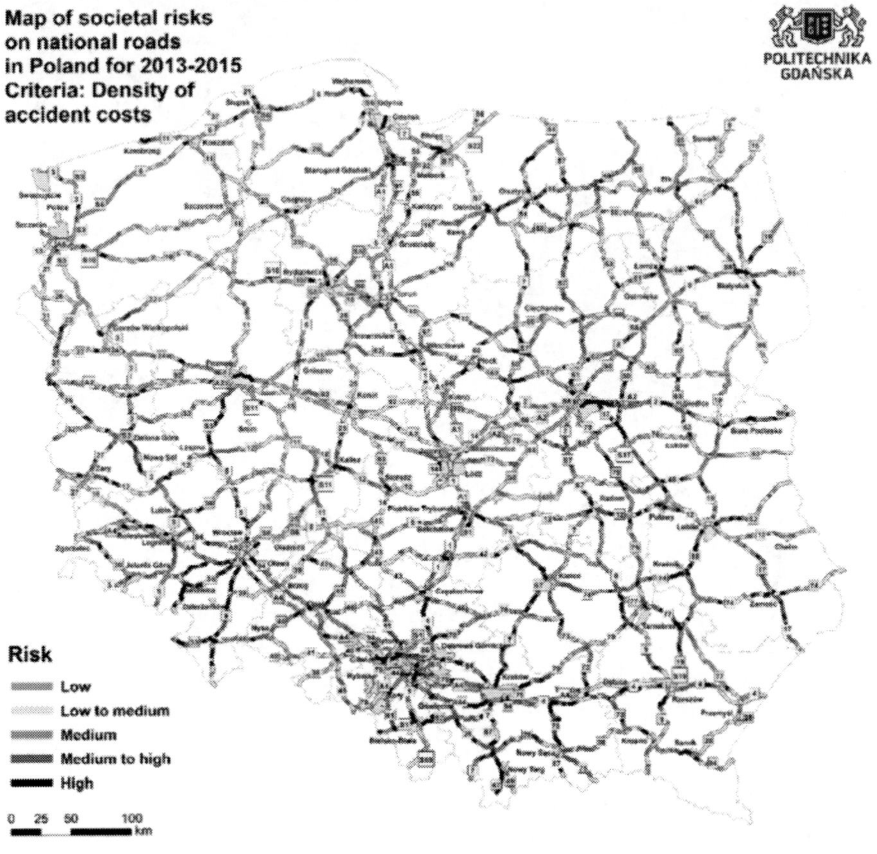

Fig. 3. Map of sections of national roads, societal risk, accident density.

Tab.1 Requirements for risk management on the network of national roads.

Road technical class	Level of acceptance of fatality accident risk		
	Acceptable risk	Tolerated risk	Not tolerated risk
	Classes of risk on a road section		
Motorway	A	B, C	D, E
Express road	A	B, C	D, E
Main road of fast traffic	A, B	C, D	E
Main road	A, B	C, D	E

6 Road safety inspection

Three types of inspection are distinguished: general (IO), detailed (ID) and special (IS). General inspections are conducted cyclically. Detailed and special inspections are conducted on road sections that have an E level of risk or as needed [8]. The road safety inspection applies to all road objects and phenomena that are important for road safety which occur on roads and in the clear zone, especially in the area of geometry, traffic layout, surrounding, visibility and road and roadside parameters in relation to the required speed (Fig. 3 – long pedestrian crossing, wrong solutions of the road intersection, trees, trees at the roadside, poor visibility).

Fig. 3 Examples of road defects.

Once the hazard assessment is completed, the next step is to determine whether the hazard will be removed or reduced to reduce the potential risk or whether hazard protection measures will be introduced. Based on this, steps should be taken to solve the problem. After a period of verification of the method (2 years from implementation), three classes of defects have been adopted (A – small risk, B – medium risk, C – high risk). At present, work is ongoing on objective road risk measures. The effects of speed and traffic volume on risk class must be defined. An outcome of the research will be a tool to help with objective ranking of road and roadside hazards identified during inspections.

7 Summary

The tools for road infrastructure safety management presented above need more research and a broader scope of application. If we can develop models of how selected factors impact road safety measures (to ensure a better selection of remedial measures), prepare road safety management methods at the operational level and introduce the Directive [2] to the network of regional roads, road safety in Poland can continue to improve.

References

1. K. Jamroz, Method of risk management in highway engineering. Metoda zarządzania ryzykiem w inżynierii drogowej (Gdansk University of Technology, Gdansk, 2011)
2. Directive of the European Parliament and of the Council of 19 November 2008 on road infrastructure safety management 2008/96/EC (Official Journal of the European Union L.319/59) (2008)
3. K. Jamroz, M. Budzyński, W. Kustra, L. Michalski, S. Gaca, Tools for road infrastructure safety management - Polish experiences, in: Transp. Res. Procedia, pp. 730–739 (2014). doi:10.1016/j.trpro.2014.10.052
4. W. Kustra, K. Jamroz, M. Budzyński, Safety PL - a support tool for Road Safety Impact Assessment, in: 6th Transp. Res. Arena, (Elsevier, 2016)
5. K. Jamroz, The impact of road network structure and mobility on the national traffic fatality rate, Procedia - Soc. Behav. Sci. **54** pp. 1370–1377 (2012). doi:10.1016/j.sbspro.2012.09.851
6. T. Szczuraszek, J. Kempa, G. Bebyn, J. Chmielewski, et al., GAMBIT Kujawsko – Pomorski, Tom II. Road safety improvement program (Diamond Books, Bydgoszcz, 2012)
7. K. Jamroz, S. Gaca, W. Kustra, L. Michalski, Road Safety Auditor's Guide - Part I Road Traffic Safety Impact Assessment of Road Infrastructure Projects, Part II Road Safety Audit - Prepared for GDDKiA (Cracow University of Technology, Gdansk University of Technology, 2011)
8. M. Budzyński, K. Jamroz, W. Kustra, S. Gaca, L. Michalski, Classification of dangerous road sections on national roads - Prepared for GDDKiA (Gdansk University of Technology, Cracow University of Technology, 2013)
9. EuroRAP, Atlas of Risks on National Roads in Poland in 2009-2011, (n.d.)
10. K. Jamroz, W. Kustra, J. Wachnicka, M. Berkowski, Risk classification methodology for selected types of road accidents on provincial roads as well as for voivodships and poviats (Ministry of Infrastructure and Development, 2015)

Road safety aspects in the management of road maintenance

Karol Markowski[1,2,*]

[1]General Directorate for National Roads and Motorways, Gdansk, Poland
[2]University of Gdansk, Faculty of Economics, Gdansk, Poland

Abstract. The last decade in Poland has been a period of intense expansion of the public roads network. During this period a spectacular increase in the number of vehicles and traffic intensities has been noticed. Considering this, organising and carrying out maintenance work is increasingly important. The paper provides a brief analysis of currently used maintenance models in Europe and Poland and their characteristics which determine the safety of both road users and workers carrying out roadworks. The historical change in the approach towards the quality of road maintenance, the evolution of the economic model and a description of how the expectations and preferences of drivers have changed are included in the paper. In addition, the paper presents specific road maintenance standards and suggestions for maintenance and road works. The summary of the paper gives a recommendation for future development directions and suggests which features of road maintenance models and standards improve safety to the highest degree, particularly in the context of the further development of the road network.

1 Introduction

The last decade has been a period of intense expansion of the public roads network in Poland. After years of delays with archaic roads that did not meet the real needs of network traffic, road construction is finally taking shape. Although this process is still ongoing, we are starting a new investment programme based on European funds. It will produce unprecedented results in the history of Polish roads. It will lead to the creation of the country's highest functional level of roads, highways and expressways. It should be assumed that these roads will carry most of the supra-regional traffic. The traffic will move with particular characteristics and specific features such as speed, travel time or behaviour of road users also in terms of their expectations of the applicable standards of living.

* Corresponding author: kmarkowski@gddkia.gov.pl

 DOI: 10.1051/matecconf/201712202009

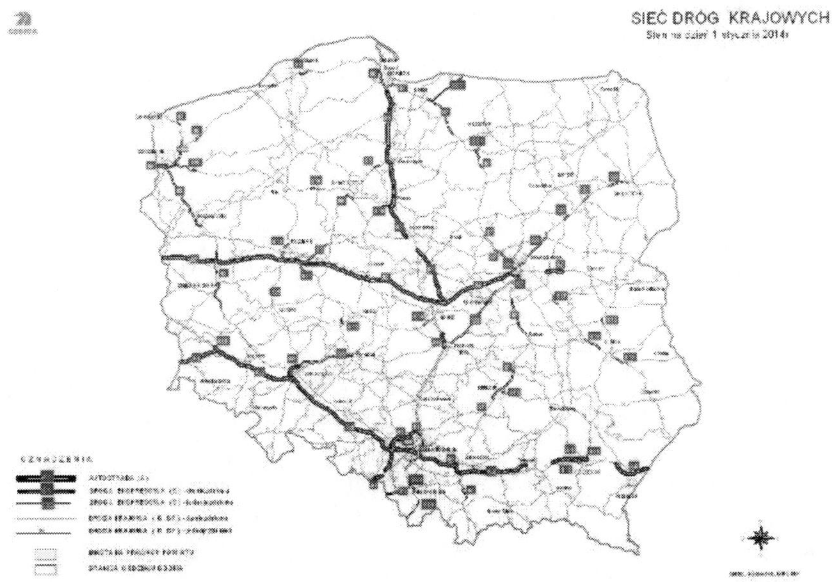

Fig. 1. The network of national roads in Poland in 2014. Source: General Directorate for National Roads and Motorways in Poland

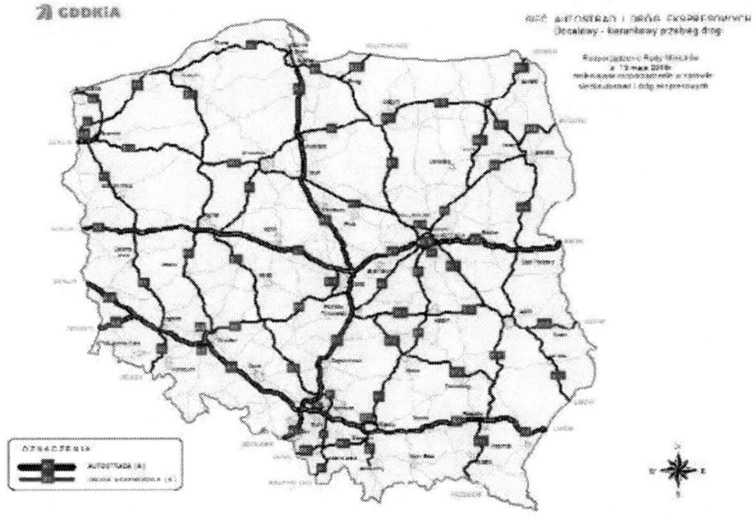

Fig. 2. The network of national roads in Poland in 2025. Source: General Directorate for National Roads and Motorways in Poland.

2 Problems of road safety in the traditional approach to road maintenance.

So far in the course of the company's operations and road safety research, particularly in the area of causes of road incidents, the problem of the quality and implementation of maintenance has never given safety a priority. It is a commonly known paradoxical claim which is that the worse the maintenance of the road is (worse quality of the roads, especially surface), the lower the risk of road accidents. This relation can be found on EuroRAP's risk maps. Of course, the paradox of this relationship is that the roads cannot be used as normal because they are in a such bad shape which forces drivers to reduce speed radically. Obviously, this cannot be treated as normal. The starting point for the analysis must be a road in good or satisfactory condition in terms of quality assessment nomenclature. It must be a road that allows drivers to drive at the speed limit safely. Only then do we realise that apart from the standard reasons for accidents caused by driver behaviour or the condition of the vehicle, there are other important reasons such as road maintenance and quality of maintenance works. Roads of the highest class, with maximum speed limits and top quality of the road must deliver high standards of maintenance to ensure safety.

The basic road safety problems for road maintenance in the traditional approach are:
- lack of homogenous standards for marking vehicles carrying out maintenance work,
- lack of homogenous standards for marking works areas,
- lack of minimum quality standards, poor maintenance of infrastructure,
- maintenance standards are not delivered, for example time of day, season, max. length of the difficulties, chaotic and uncoordinated outsourcing of maintenance with no relation to the real needs,
- lack of operating standards in an emergency situation on the road,
- lack of cooperation standards with emergency services,
- lack of assessment indicators to meet the standards, especially in terms of quality,
- lack of uniform rules to control and determine the consequences of failure to meet the standard.

Maintenance work carried out by different managers (sometimes even within a single organization) often differs not only in the scope or frequency, but also the form, method of delivery and method of securing the work zone. These inconsistencies can adversely affect safety [1].

3 The experience of European countries

The aspect of road safety in the implementation of maintenance works is present in the practice of all road managers in Europe. No matter which model of maintenance has been accepted, countries have developed their characteristic approach to the problem.

Table 1. The maintenance management system in England [2].

The Board of Roads	The Highways Agency
Model of maintenance	Comprehensive, in PBC system
Standards	Defined, catalogued (BVPI catalogue), taking into account the expectations of road users (Citizens Charter)
The quality of maintenance	At the level of expectations

Table 2. The maintenance management system in Germany [2].

The Board of Roads	
Model of maintenance	Traditional, implemented by the federal states, the Länder
Standards	At the internal level, enforced on the basis of indicators
The quality of maintenance	High

Table 3. The maintenance management system in Austria [2].

The Board of Roads	ASFINAG (commercial company)
Model of maintenance	Professional, comprehensive
Standards	Defined also in the busy belts including maximum delay of the timing of traffic,
The quality of maintenance	High

4 Historical changes in maintenance model, taking into account aspects of road safety.

In Poland, the transformation of the management model of road maintenance and the evolution of the approach to road safety went hand in hand with general changes in the socio-economic transformation of the political system in 1989. In a free market economy the economic effectiveness of the approaches has changed considerably. Effectiveness of the works in reducing road incidents also came to the fore. The importance of these problems was increasing in subsequent years in proportion to the development of the road network and road traffic.

The unquestionable leader of these changes was from the beginning, and still is the largest manager of public roads in Poland, which is the General Directorate for National Roads and Motorways. The General Directorate for National Roads and Motorways is thought to be the unquestionable leader of these changes. The company manages the whole country network of more than 19 000 km of national roads of different classes and different characteristics. This network, since the establishment of GDDKiA until now, has been managed on the basis of the evolving model of maintenance, using the strength of their own forces and resources, through models of commissioning work at different levels of aggregation, finally ordering a comprehensive maintenance model called “Keep Up the Standard”. The process proceeded in a planned and systematic way.

Maintenance management model has evolved in the following stages [3]:
- Stage 0, 1989 - 1999 - partly traditional model / partly commissioned works,
- Stage I, 1999 - 2004 - commissioned distributed works model,
- Stage II, 2004 - 2011 - commissioned cumulative works model,
- Stage III, 2011 - 2015 - “quasi standard” model,
- Stage IV, from 2016 - “Keep Up the Standard” model.

4.1 Stage 0 (1989 – 1999) – partly traditional model

The model implemented in the period before the GDDKiA was established within the framework of its legal predecessors, took place in the period of political transformation i.e. from 1989. Maintenance was carried out by in-house employees with some commissioning of more specialized jobs. The road manager had their own employees and some equipment, trucks and vans. Materials for repairs and maintenance (signs, posts, concrete elements, materials for minor repairs, etc.) were bought by the road manager.

Features of the model:
- GDDKiA road workers: in-house/full-time employment,
- equipment: own,
- materials: own
- outsourcing: only for specialized works,
- lack of uniform standards of road safety,

- lack of a system to control the performance of your own work (ad hoc audit)
- lack of indicators to assess the effectiveness of road safety solutions.

4.2 Stage I (1999 – 2004) - commissioned distributed works model

When the administrative changes in the country and the creation of GDDKiA took place, the manager continued efforts to merge the different types of outsourced works, which mostly led to a passive human resources policy at the level of workers and to minimizing the number of their existing equipment. Outsourced maintenance work began in larger packages along with the determination of the manner of implementation.

Features of the model:
- GDDKiA workers: minimum number,
- GDDKiA equipment: minimum level,
- outsourcing: outsourced by types of works with defined methods of implementation,
- lack of uniform road safety standards,
- control system implemented by GDDKiA employees towards third parties performed only by checking compliance with the contract,
- lack of indicators to assess the effectiveness of road safety solutions.

4.3 Stage II (2004 – 2011) - commissioned cumulative works model

From 2004 GDDKiA in its new orders introduced specific scopes depending on the groups of works to which they belonged. In particular, works were aggregated into groups: winter maintenance, routine maintenance of roads, bridges routine maintenance. The agreement assumed specific scopes of road maintenance, repetitive work was valued individually, separately for each item.

Features of the model:
- GDDKiA workers: none
- equipment: none
- materials: property of the contractor
- outsourcing: outsourced by assortments with defined methods of implementation,
- lack of uniform road safety standards,
- control system implemented by GDDKiA employees towards third parties performed only by checking compliance with the contract,
- lack of indicators to assess the effectiveness of road safety solutions.

4.4 Stage III (2011 – 2015) - "quasi standard" model

The model of commissioned works was to include quality indicators. Prior to that in 2015, the Gdansk branch of GDDKiA introduced an intermediate model. It was agreed that previous framework contracts would bring together all the specific works with the traditional division of current road maintenance, winter road maintenance and current bridge maintenance. One contractor did all the work in the entire region. Fixed work schedules were introduced with first standards for work and quality of maintenance. The contractor was informed about the requirements regarding staff, equipment and facilities. Works became standardised. All regulations were fully harmonized across the road network irrespective of the Region. For the whole road network the Department "Order Description" made sure that the same principles of carrying out work were followed. The only difference was in sections of roads and the scope of work. Uniformed rules for commissioning, inspection and acceptance were also introduced, among others the so-called "matrix of control" was introduced.

Features of the model:
- GDDKiA workers: none
- equipment: none
- materials: none
- outsourcing: clustered in groups of 3 assortments,
- the first uniform standards for the whole region were introduced,
- control system implemented by GDDKiA employees towards third parties performed only by checking compliance with the contract,
- lack of indicators to assess the effectiveness of road safety solutions.

4.5 Stage IV (from 2010 to 2016) - "Keep Up the Standard" model.

In parallel, from 2010, work began on developing the principles of quality indicator-based contracts. Initially the system was used on selected sections of expressways but has been extended to cover more than one road section. A pilot model "Keep Up the Standard" was introduced and from 1 January 2016 the road maintenance system now covers the entire network of national roads across the region. The first solution was applied by GDDKiA Branch in Gdansk.

Features of the model:
- GDDKiA workers: minimum number,
- GDDKiA equipment: minimum level,
- outsourcing: one contract including all the maintenance work, using a flat-rate system contracted by assortments with defined methods of implementation,
- uniform standards for road safety including:
  - standards for marking equipment and personnel,
  - standards for conducting ongoing monitoring of road safety (road safety patrols)
  - standards for work signage,
  - standards of safe work,
  - standard methods for carrying out the work, the time of day, day of week
  - standards for achieving readiness,
  - standards of cooperation with emergency services,
  - standards of the technical condition of road elements that may have an impact on road safety,
- indicator system for evaluating the effectiveness of road safety solutions,
- professional system of control of standards carried out by employees of GDDKiA to check external companies,
- model with features similar to systems used in the countries of Western Europe.

Table 4 Maintenance management system "Keep the Standard" in Poland.

The Board of Roads	GDDKiA
Model of maintenance	Complex "Keep Up the Standard"
Standards	Specific, consistent for the whole network
The quality of maintenance	High, at the level of expectations

5 Conclusions and recommendations.

The dynamic development of the road network, expansion of the road system of the highest category dedicated to high speeds and increased traffic flows, as well as the expectations of the drivers themselves mean that targeted solutions for infrastructure management will be required with road safety as a priority. Such models already exist in other European countries, and although they differ from each other in the details they have much in common. It is clear that they take into account the impact of the quality of maintenance and delivery of maintenance works to reduce the risk of road accidents. While there are many different approaches to the same model of road infrastructure management, the recommended activities will be designed to improve road safety. They are:

- improving the efficiency of operations and, consequently, the quality of road maintenance,
- the introduction of standardized ways of implementing maintenance works,
- using tools and processes to support maintenance, such as: professional road information systems, ITS systems, emergency procedures.

There is no doubt that among the currently used models of maintenance, these objectives will be achieved fairly easily by using a system of indicators such as the model of area maintenance or "Keep Up the Standard" system. It is important to note, however, that the final shape of the model both in terms of quality standards and methods of road works should be preceded by research to be conducted along with the development of infrastructure, taking into account the experience of other countries and respect for regional conditions.

References

1. E. Marciszewska, J. Pieriegud, Benchmarking and Best Practices in Transport Sector (SGH, Warsaw, 2009)
2. J.Archutowska, J.Pieriegud, Effectiveness of the National Roads Maintenance Management in Poland (Ernst&Young, Warsaw, 2012)
3. K. Markowski, "Keep Up the Standard" in terms of area on the national roads of the Pomeranian province administered by GDDKiA Gdańsk Branch (Warsaw, 2010)

Speed management on local government managed roads – research, recommendations and guidelines

Marcin Budzyński [1,*], *Stanisław* Gaca[2], and *Mariusz* Kieć[2]

[1]Gdansk University of Technology, Faculty of Civil and Environmental Engineering, Gdansk, Poland
[2]Cracow University of Technology, Faculty of Civil Engineering, Cracow, Poland

Abstract. Commissioned by the National Road Safety Council Secretariat, the project "Guidelines for speed management on local government managed roads" studied car driver behaviour when subjected to selected speed management measures such as: local speed restrictions, surveillance, traffic calming and restricted speed areas. In addition, analyses were conducted on the impact of selected measures on the level of road safety. The behaviour was assessed by studying the changes in speed parameters (statistical characteristics) for the particular speed management measures or their absence. The road safety level was measured by comparing the level before and after the implementation of the particular speed management measure, taking trends into account. The paper presents the results of the research, along with recommendations for speed management guidelines.

1 Introduction

Speed management is a set of measures designed to set reasonable speed limits and influence the drivers' choice of speed. This can be achieved through urban planning, infrastructure and traffic layout, enforcement, education and advanced technologies. The basic goal of speed management is to achieve a state of traffic where vehicle speeds are adjusted to the road and traffic making the speeds potentially safe. If properly managed, speed control can help reduce traffic noise and air pollution.

There is evidence from research and practical experience that road safety can be significantly improved through consistent speed management, which includes road engineering measures. There are a number of engineering solutions that can help to reduce vehicle speeds such as road infrastructure design that helps drivers to choose the right speed, physical means to control driving speed and reasonable speed limits. Some of the measures are not very restrictive and leave the decisions to drivers (how willing they are to accept a restriction) and local road and traffic conditions. In this case the actual benefits from changing speeds and better road safety depend on social and cultural factors. As a consequence, overreliance on the results of international speed management research may not offer the same effects in Central and Eastern European countries. This has prompted new research into how different speed management measures affect driver behaviour and road safety. The paper describes the effectiveness of selected speed management measures and

* Corresponding author: mbudz@pg.gda.pl

the effect they have on road safety. To that end the power model was used. It was calibrated based on research results.

2 The effects of speed management on road safety

The effectiveness of speed management measures and how they contribute to better road safety can be analysed using direct (road incident data – collisions and accidents) or surrogate measures that describe the potential hazards. One way to understand the effects of speed management on road safety is to estimate how a specific safety measure has changed. The rate is a quotient of the assumed change in safety on a treated section and the mean value of that same measure on an untreated control section. This is referred to in the Highway Safety Manual as CMF (Crash Modification Factor) which is the basic rate with which to evaluate how different road safety treatments change road safety [1].

The effects of a speed management measure on road safety can also be estimated using surrogate rates such as change in vehicle speed as a result of the measure. There is evidence that speed can be used as a surrogate road safety rate known from international and Polish research. The statistical relation between speed and road safety is logical and has been demonstrated many times before [2–7]. To estimate how a change in vehicle speed triggers a change in road safety, we can use the "power model" [4]. It helps to predict accidents and accident casualties based on the difference in mean speed "before" and "after" a measure has been applied. To this end, the equation below is used:

$$W_1 = (V_1/V_0)^a * W_0 \tag{1}$$

where:
- W_0 – selected road safety measure in the observation period prior to treatment,
- W_1 – selected road safety measure in the same observation period after the treatment,
- V_0 – average speed prior to treatment [km/h],
- V_1 – average speed after the treatment [km/h],
- a – model parameter with the value based on literature or set individually.

A key assumption when using the relation (1) is that during the "before" and "after" analysis, none of the other road safety determinants change, except speed. Research described in [4] was used to determine the following values of a in equation (1) (Table 1).

Table 1. Total time from the first look until the target is passed and Time during which driver's gaze is fixed on target (ms) in 4 analysed situations [4].

Type of accident	Rural roads		Urban roads	
	a parameter	Confidence level 95%	a parameter	Confidence level 95%
Fatality accident	4.1	2.9 ÷ 5.3	2.6	0.3 ÷ 4.9
Fatality and serious injury accident	2.6	-2.7 ÷ 7.9	1.5	0.9 ÷ 2.1
Total accidents involving casualties	1.6	0.9 ÷ 2.3	1.2	0.7 ÷ 1.7

If we assume the above a values, in the case of urban roads, a speed limit reduction from 60 km/h to 50 km/h, for the average baseline speed of 60 km/h and a speed reduction of 2.5 km/h, fatality accidents would drop by 10.5% with total accidents going down by 5%. With improved speed limit compliance (e.g. through enforcement or physical means of traffic calming) and a real reduction in average speed by 5 km/h, fatality accidents could drop by 21% and total accidents by 10%.

Figure 1 shows an example of how the VTI power model can be used to estimate the effects of changing the mean speed on a specific road section on the percentage of accidents of different degrees of severity. The figure shows how a change in mean speed on a road section (by 3 km/h, 6 km/h, 9 km/h, 12 km/h or 15 km/h) can potentially improve road safety relative to the mean speed before the change.

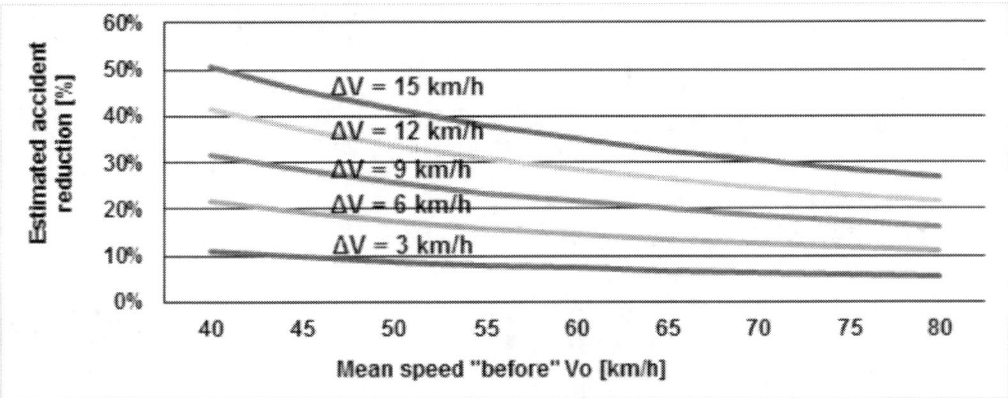

Fig. 1. Estimated reduction in fatal and serious injury accidents in relation to the initial mean speed and speed reduction ΔV – roads in built-up areas [8]

As well as the less restrictive measures, physical traffic calming measures are also used. These are much more effective and include speed bumps and raised surfaces. International research [9] suggests that speed bumps are highly effective and help to reduce vehicle speeds by about 32 km/h and about 27 km/h when speed tables are applied. Raised pedestrian crossings used as an obstacle produce results similar to speed bumps with average speed reductions by 4.0 ÷ 6.5 km/h. Please note, however, that while speed bumps reduce speed locally, speed increases in between the bumps which increases emissions (accelerating and braking is more frequent). This is why it makes more sense to use comprehensive traffic calming measures with speed reductions distributed evenly on a designated road section. Mini roundabouts and small roundabouts in place of regular junctions reduce average speed by 36 km/h and 54 km/h respectively.

The results make it very clear that speed management is most effective when physical traffic calming measures are used. Despite that, before any such treatment, an analysis should be conducted to look at the effectiveness of other measures such as active speed limit signs, intensive enforcement, adding or removing lines separating traffic lanes and allowing vehicles to be parked along the road.

3 Research and results

Research into the effectiveness of speed management measures included accident and casualty analyses "before" and "after" a treatment (in locations where the date of a treatment was known). If the date of a treatment was not known, the trends in accident numbers were analysed. Surrogate safety measures were studied too, i.e. vehicle speeds. The research covered the following cases:

- local speed limits (Figure 2a and b) – 44 locations for accident analysis and 27 locations for speed analysis
- TEMPO 20 residential area (Figure 2c) – 10 locations for accident analysis and 8 locations for speed analysis
- TEMPO 30 area speed limit (Figure 2d) – 35 locations for accident analysis and 12 locations for speed analysis
- sections with mild traffic calming measures (change in vehicle trajectory while providing good passage for heavy vehicles, Figure 2e) – 33 locations for accident analysis and 29 locations for speed analysis

- sections with additional signs to inform of a possible speed control on the section (Figure 2f) – 30 locations for accident analysis and 12 locations for speed analysis.

Fig. 2. Examples of analysed speed management measures [8].

3.1 Speed analysis

Due to the conditions on site, speed was measured using three measurement methods:
- manual (observations in the field in subsequent measurement cross-sections or analysis of recorded video footage),
- automatic (pneumatic detectors)
- a combination of manual and automatic measurements.

Manual measurements helped to determine vehicle speeds in free flow traffic only, split into two types of vehicles (light and heavy vehicles). The advantage of the method is that it is done without the drivers noticing. This method was used in Tempo 20 and Tempo 30 zones and in the case of local speed limits.

Automatic measurements helped to ensure that speed was recorded very accurately while differentiating between vehicle speeds in the entire traffic flow and in free flow traffic. The test devices were able to identify vehicle types. This method was used on sections with local speed limits, mild traffic calming measures and on speed control sections.

Speed was measured on sections with speed management measures and on control sections (with no speed management measures). In the case of area speed limits the control sections were selected on streets with Tempo 20 and Tempo 30 characteristics. Mild traffic calming measures were represented by results from sections of a similar characteristic (roadside, cross-section, accessibility). Untreated sections of the same road were also considered. Local speed limits and section control were represented by the following cross-sections: app. 150 m before the sign (as a control section), at the sign and behind it (app. 150 m). Control sections representing local speed limits had similar geometrical parameters. The results show that:

- local speed limits help to reduce average speed significantly compared to the control cross-section (no speed restrictions). The reduction depends on the limit and changes from app. 6.8% to 14.5% on treated sections;
- temporary speed enforcement signs help to reduce speed for longer compared to speed cameras. This measure reduces average speed by about 15.8% compared to the control section;
- mild traffic calming measures help to reduce average speed by about 10.3% with vehicle speeds becoming similar;
- while Tempo 30 zones help to reduce speed by app. 16.3% compared to control sections, the spread of speeds was quite strong with speeds depending on the local road conditions and environment;
- residential area (Tempo 20) results show a speed reduction by about 13.4% in free flow traffic compared to control sections. Just as with Tempo 30 zones, the spread of speeds was significant and depended on local road conditions. Please note the high percentage of drivers going above the speed limit in residential areas.

Fig.3 shows examples of test results when the speed limit was changed to 70 km/h in a non-built-up area and a comprehensive traffic calming treatment was introduced in a small town (for free-flow). Results of the analysis show that a local speed limit of 70 km/h on average:

- reduces mean speed by 9.3 km/h for free-flowing vehicles; compared to the control cross-section the reduction is by 11%;
- reduces V85 by 10% for free-flow traffic,
- the averaged standard deviation changes compared to the control cross-section which suggests that the local speed limit has not improved traffic homogeneity.

Analysis of a group of road sections with calmed traffic in small towns shows that this treatment:

- reduces average speed by 2.8 km/h for free-flow; this is a 5% reduction compared to the control cross-section;
- reduces V85 speed by 7% for free-flow but the effect differs from site to site and ranges from an increase of 2% to a decrease of 27%; this strong variation may be the result of the traffic calming measures used;
- reduces standard deviation by 1.7 km/h (16%) for free-flow compared to the control cross-section which shows that comprehensive traffic calming treatment on roads passing through small towns improves traffic homogeneity.

DOI: 10.1051/matecconf/201712203001

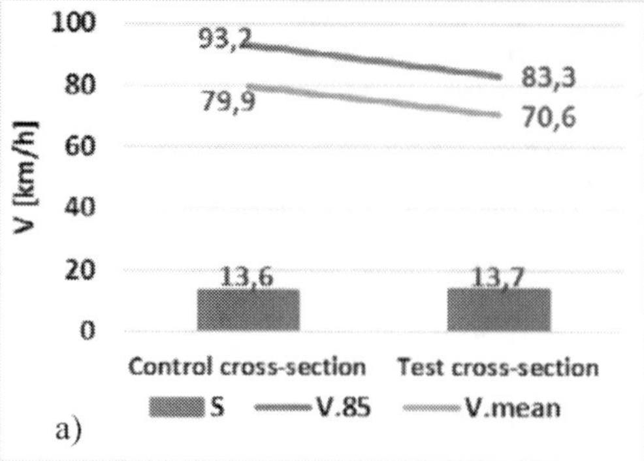

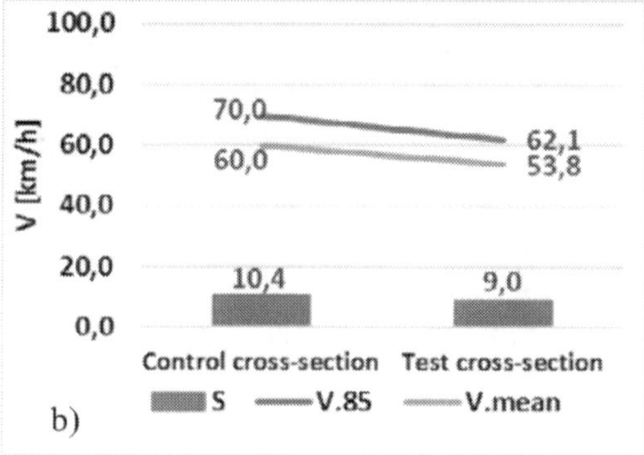

Fig. 3. Average speed reduction in free –flow (a - Local speed limit –70 km/h, b - Comprehensive traffic calming in a small town [8].

3.2 Accident analysis

With limited data availability, the analysis of speed management effectiveness was simplified. The annual number of accidents and casualties was determined for all the locations under analysis from 2006 to 2014. The next step was to choose those locations whose date of treatment was known and fell within the period 2007 – 2013. Once selected, the locations were then analysed for their annual average road safety rates "before" and "after" the treatment. This produced the quotients of the "after"/"before" measures under estimation for the specific speed management measures. The lower the value of the quotient, the higher the effectiveness of the measure. This is a typical approach to "before" and "after" effectiveness evaluations. Because accidents may change regardless of the treatment, an additional comparison was made of the quotients to general accident trends in the regions under analysis. The effectiveness of the measures was estimated by comparing the "after" quotients to the "before" situations for the treated sites and control sections. The bigger the difference between treated site and control section values, the more effective the treatment is (under the assumption that test site values are lower than those in control areas).

Table 2 lists the effectiveness indicators of the speed management measures and the resulting accident and casualty reduction. Because residential areas have a low number of accidents (Tempo 20), those test sites were analysed together with Tempo 30 zones.

Table 2. Analysis of the effects of speed management on road safety [8].

Measure	Accident reduction	Fatality and serious injury reduction
Local speed restriction	22% (26% - speed-related)	15% (18% - speed-related)
Intensive enforcement	45% (52% - speed-related)	48% (56% - speed-related)
Traffic calming on roads in small towns	No statistically significant impact	No statistically significant impact
Traffic calming on roads in cities	17% (25% - speed-related)	14%
Tempo 20, Tempo 30	No statistically significant impact	12% (30% - speed-related)

The results show that speed enforcement (speed control section signs) and area speed limits are effective in improving road safety. It is important to note, however, that the results are based on a relatively small sample and more research is required.

With the evaluation of how different speed management measures affect road safety on Poland's lower class roads, it is clear that the effects differ from measure to measure. Using the results of "before" and "after" comparisons and taking account of the general road safety trends, the following are the initial evaluations of how the particular measures affect road safety:

- no positive effects were established in Tempo 20 and Tempo 30 zones on accident reduction. Fatalities and serious injuries, however, are down – the average reduction rate being 30% (accidents involving speed);
- contrary to expectations, mild traffic calming measures had no positive effect on accident reduction or casualty reduction;
- sections of roads with local speed limits have seen a positive effect of the restrictions with fewer accidents (a 22% reduction) and serious injuries and deaths (a 15% reduction);
- sections with speed limit enforcement measures have seen very positive effects both in terms of accidents (a 45% reduction) and serious injuries and deaths (a 48% reduction).

3.3 Power model calibration

The next step of the analysis was to calibrate the "a" parameter for the "power model" given in equation (1) using the speed results and accident analysis. The factors calculated include speed-related accidents, severity of all accidents and speed-related accidents which should have a strong relation to speed. Table 3 gives the parameter's values for the measures analysed. The "a" parameter for local speed limits has not been evaluated because of insufficient accident sample for each value of a speed limit.

Table 3. The "a" parameter in the "power model" equation based on authors' research

Speed management measure	**Power model's *a* parameter**		
	Speed-related accidents	Total serious injuries and fatalities	Speed-related serious injuries and fatalities
Speed limit	-	-	-
Residential area (TEMPO 20)	3.2	3.0	5.2
TEMPO 30	2.6	2.4	4.2
Mild traffic calming measures	2.6	0.0	-1.0
Speed control sections	6.0	4.9	6.3

4 Conclusion

The research into the effects of speed management measures on road safety shows that the measures have a varying effect on safety when measured with direct rates such as accidents and accident severity and with indirect rates such as vehicle speed.

The studies confirmed a highly beneficial effect of speed intensive surveillance as a measure to improve road safety. Local speed restrictions (B-33 signs) do not generally bring the desired speed, but significantly affect speed reduction and decrease the number of accidents and casualties. Recorded speed reduction in the local restriction areas by 9.3 ÷ 11.9 km/h (11.6% ÷ 14.6%) means a potential decrease in the number of accidents with fatalities and serious injuries by approx. 27 ÷ 34% – on roads outside built-up areas. Traffic calming on through road sections caused a slight reduction in speed but did not have a statistically significant effect on road safety improvement – this may result from the specifics of the applied traffic calming measures

Traffic calming on street sections in towns did not result in average speed reduction. There was no statistically significant impact on accident reduction, but a decrease in the number of fatalities and serious injuries was observed – this surprising result of the research may be associated with the specifics of the applied traffic calming measures. In the 30 km/h and 20km/h zones a decrease was registered in fatality and serious injury accidents. While "mild" calming measures are not very effective, physical measures such as speed bumps and raised junctions proved to be very effective.

The results can be used to build "power models" that help to understand the effectiveness of speed management measures. This will ensure that proposed treatments will offer the best effectiveness for a given set of conditions.

References

1. Highway Safety Manual (HSM) (AASHTO, Washington DC, 2010)
2. M.H. Cameron, R. Elvik, Nilsson's Power Model connecting speed and road trauma: Applicability by road type and alternative models for urban roads, Accid. Anal. Prev. **42** pp. 1908–1915 (2010)
3. S. Gaca, Study of vehicle speed and its impact on road safety (Cracow University of Technology, 2002)
4. R. Elvik, A. Hoye, T. Vaa, M. Sorensen, The handbook of road safety measures. The second edition (Emerald Group Publishing, 2009)
5. Austroads Ltd., Infrastructure/speed limit relationship in relation to road safety outcomes, AP-T141/10 (2010)
6. S. Gaca, M. Kieć, Vehicle responsiveness tests to change speed limits on built-up areas, Urban Reg. Transp. **12** (2005)
7. K. Jamroz, W. Kustra, M. Budzyński, J. Żukowska, Pedestrian protection, speed enforcement and road network structure as the key action for implementing Poland's Vision Zero, Transp. Res. Procedia. **14** pp. 3905–3914 (2016). doi:10.1016/j.trpro.2016.05.479
8. S. Gaca, M. Budzyński, M. Kieć, Guidelines for speed management on local government managed roads (Secretariat of the National Road Safety Council, 2016)
9. Federal Highway Administration, Speed management: A manual for local rural roads owners, FHWA-SA-12-027 (U.S. Department of Transportation, 2012)

Research on the parameters of light emitting advertising media

Marcin Chrzanowicz[1,*], and *Piotr* Tomczuk[2]

[1]Warsaw University of Technology, Faculty of Electrical Engineering, Warsaw, Poland
[2]Warsaw University of Technology, Faculty of Transport, Warsaw, Poland

Abstract. This article presents preliminary test procedures for measuring the light performance of advertising media related to the way that the light beam affects the driver. A basic distribution of advertising in terms of light emission and estimated luminance levels that can cause blinding. Finally, the results of the diminishing of the luminance of the surface; examples of luminous ad types and identifies their impact on visibility in terms of the possibility of blinding that results in the deterioration of road safety (especially at night).

1 Introduction

The problem of the assessment of the impact of advertising media located in the vicinity of roads on motorists can be considered in a number of ways. The factors which can draw the driver's attention to the advertising medium have many causes. You can classify them as: aspects of the content, location, geometry, psychological and lighting. Each of the foregoing factors may significantly affect the conditions and hinder the driver's vision.

The study of the effects of advertising media located in the vicinity of roads within the driver's view [1,2] testify to the existence of a number of research problems related to lighting performance. The research carried in Poland [3,4] and a number of other countries [5–7] shows the need for the use of lighting parameters such as luminance and contrast [8,9] in the assessment of the impact of advertising on drivers. The lighting parameters that can be analysed in the context of advertising light are the advertising's luminance area, the luminance contrast between the advertisement area and the background, the distribution of the colours of the light, the illuminance on the plane of the driver's eyes produced by the advertising area and how the light is emitted (continuous or variable, directional or diffuse). In addition, you can analyse geometric parameters such as the angular size of the advert and the location of the field of vision. Advertising media is often characterized by variable luminance on its surface; this is particularly evident in the case of advertising with an external light source. The maximum value is obtained by examining the surface of the advertising perpendicular to the geometric axis of advertising media. However, with the increase in viewing angle value decrease is therefore an important issue is to determine the actual levels of luminance advertisements [2,7] of the lines of sight of the driver.

The luminance contrast with the background of the ad space is the ratio of the luminance difference between the observed object and the luminance of the background. Large values of luminance contrast can blind drivers [12]. The possibility of misleading the driver is a

* Corresponding author: marcin@zkue.ime.pw.edu.pl

DOI: 10.1051/matecconf/201712203002

dangerous aspect of road safety. In the case of advertising light there is such a risk as a result of similar lighting values and emission colour from the advertising carrier and the indicator light. In order to determine these parameters and neutralize the threat it is necessary to study advertising in terms of luminance and chromaticity and comparison with the colours of regular traffic light.

Driver distraction can be caused by the excessive brightness of advertising media, uneven distribution of luminance, high contrast and high luminance variation in time (variable message advertising).

2 Division of advertising in terms of the direction of the emission of the light beam

The introduction of the split of advertising is necessary because of the different visual impressions caused by different light distribution. The advertisement area can emit a light beam in a perfectly distributed (Lambert reflection) and mixed way. The most disturbing LED directional advertising light distributions are shown in Figure 1.

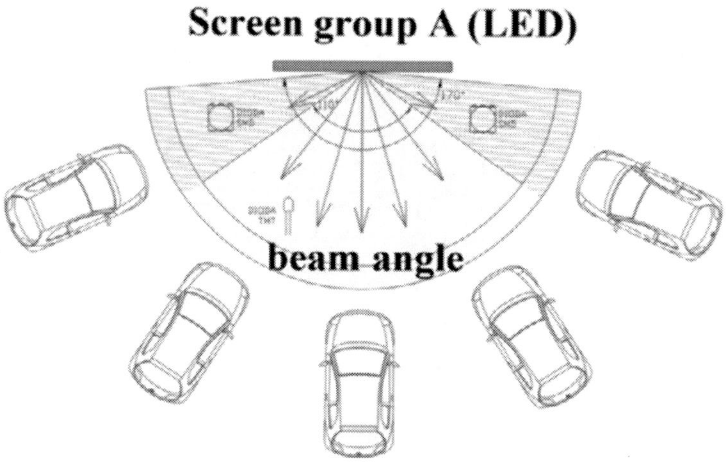

Fig. 1. Character of the emission light emitted from LED advertising.

The presented character of the distribution means that the impact of the media is highly dependent on the angle of observation, and thus at certain geometrical settings the advertising will increase the impact of distractions and at others it does not interfere. Maximum glare occurs in a direction perpendicular to the carrier. As a result of the study, the field analyses of the advertising suggested the division shown in Figure 2.

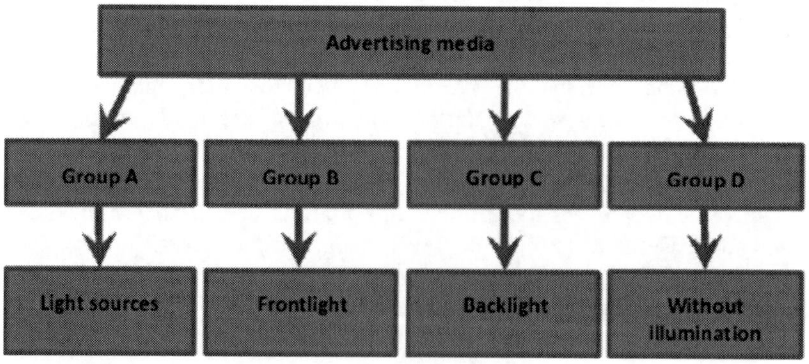

Fig. 2. The division of advertising media in terms of light emission.

Group A included media whose surface is also a source of light where the emission beam is highly directional. Group B and C are carrier surfaces which are illuminated by additional fittings (from the front and the rear of the vehicle respectively), and thus their reflection characteristics is similar to Lambert reflection light scattering (ideal dispersion). Group D is the media that do not emit light. It should be noted that the division is presented as a preliminary concept which will be expanded with of indirect and mixed variants.

3 Initial test procedures with examples of test results of lighting parameters

The carriers of variable content in which the surface luminance changes in a dynamic way are a separate issue. The distribution of light intensity affecting the eyes of a driver in the vicinity of the advertisement is presented in Figure 3.

Registering presented the course requires a light meter with very fast sampling (approximately 25 Hz). A large gradient change of lighting parameters can also distract, and the scope of changes force the team investigating to find the worst-case (to determine the maximum gradient variations in lighting on the driver's eyes).

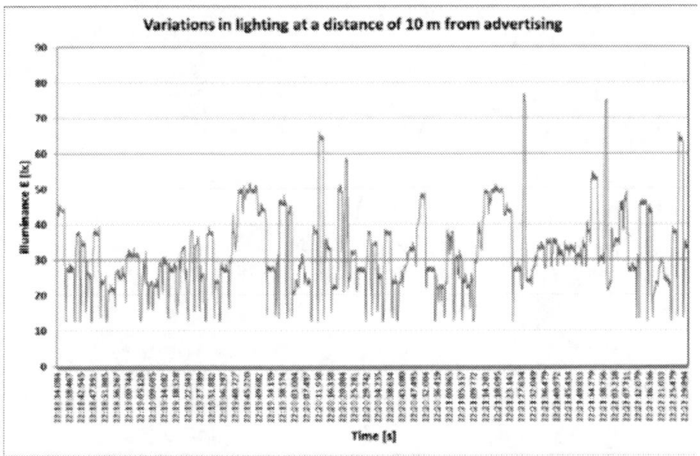

Fig. 3. Distribution of advertising media due to the emission of light.

Considering the luminance distribution of advertising media it is necessary to develop a measurement procedure which explicitly and repeatedly verifies the worst case emission of the light beam. The first prerequisite is the measurement of the direction with the largest emission of light (in the driver's field of vision). It is usually the direction perpendicular to the support (depending on the photometric solid). The second condition is to define the maximum size of the measuring field and a minimum amount of measurement areas. In order to search for the optimal values the measure of the luminance of the surface of advertising as a function of the size of the measuring field were analysed. They were chosen from the group ad A that has a surface light source. The purpose of the measurement was to find the maximum value. Taken measurement of the luminance distribution meter matrix CCD whole carrier surface with a resolution of approximately 320 thousand points, treating it as a study model (Fig. 6). Then analysed variables size measurement fields and their impact on the value of the maximum luminance (Fig. 4 and 5). The measurement data are summarized in Table 1, together with an analysis of the error with respect to the calibration measurement (meter matrix).

Fig. 4. The search for value Lmax [cd / m2] in one field of measurement (126000 pix).

Fig. 5. The search for value Lmax [cd / m2] for 15 measurement fields.

Fig. 6. The search for value Lmax [cd/m^2] using the meter matrix (318750 points - reference value).

Tab. 1. Investigating the maximum of luminance as a function of the size of the measuring field.

No.	Maximum luminance Lmax [cd/m2]	Measurement error [%]	Number of picture [-]	Comments
1	132.1	69.2	6	1 measurement field 126,000 pix
2	286.2	33.3	5	15 fields (5x3)
3	383.5	10.62	-	325 fields (25x13)
4	414.8	3.34	-	1250 fields (50x25)
5	429.1	-	4	1 measurement field (750x425) 318,750 pix

Analysing the results of the measurement of the luminance with the maximum greatest accuracy is achieved at the highest level of the discretization of the distribution surface. This follows from the fact that the gauge size of the luminance averages the maximum in the measurement. With a large field of measurement, there is a good chance that I'll be in the bright areas, but also heavily blacked out, which generally Fake it worth the read. When creating the procedures the technical capabilities of existing measures and technical conditions of measurement should be considered. The optimum division was at least 10 boxes horizontally and 10 vertically. Depending on the class of the meter (the size of the angular measurement field) it was necessary to adjust the distance measurement to obtain a box at the right surface.

In addition, measurements of the luminance distribution were taken of the surface of the selected media supports from all four representative groups. A measure matrix was used, as it was the most accurate measuring tool. The results are shown in Table 2 and Fig. 7 (maximum value and average luminance of the maximum of four similar supports).

Tab. 2 Measurements of Lmax selected ads of four representative groups.

Group	Maximum of luminance L_{max} [cd/m²]	Average of luminance $L_{maxšr}$ [cd/m²]	1 advertise-ment L_{max} [cd/m²]	2 advertise-ment L_{max} [cd/m²]	3 advertise-ment L_{max} [cd/m²]	4 advertise-ment L_{max} [cd/m²]
A	1087	1024	987	1005	1087	1018
B	103	78	60	84	103	64
C	278	257	245	234	278	271
D	10	4	8	11	10	12

The results obtained indicate that the highest luminance values are carriers of group A and C, and they may cause the greatest driver distraction.

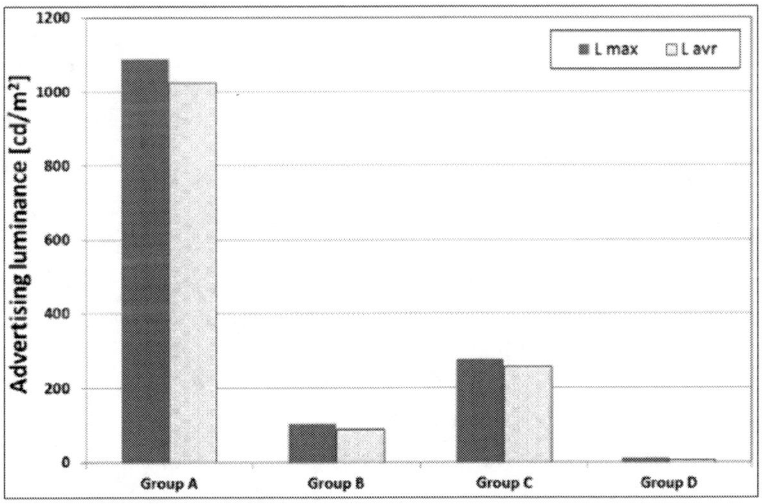

Fig. 7. Comparison of the maximum and average of luminance of representative ad groups.

4 Summary

It is very important to undertake research towards establishing the acceptable levels of the performance of light beams emitted by ads. Each distraction of driver's attention visible on the road should have a very specific impact which should not significantly worsen the conditions of the driver's vision. In addition to specifying the limit value of the maximum luminance it is necessary to establish an unambiguous measurement procedure, so as to prevent undervaluation. The most important thing is to maintain the correct directions of measurement and the optimal size of the advertising media measuring field. The ad group A (LED, OLED) requires particular attention when creating the legal requirements and procedures. Further research will also be required to determine the impact of the reflective properties of the materials used on advertising media surfaces and the way the light beam affects the driver's visual competence which has a direct impact on road safety.

Acknowledgements

The issues presented in this article are the subject of the research and of analyses in the project carried out by the Department of Transport of the Warsaw Technical University entitled: Influence of Advertisements on the Security Level of Road Traffic, agreement number: DZP/RID-I-33/4/NCBR/ 2016 (Development of Road Innovation), financed by the National Centre of Research and Development and the General Management of National Roads and Motorways.

References

1. D. Beijer, Driver distraction due to roadside advertising (University of Toronto, Department of Mechanical and Industrial Engineering., 2002)
2. T. Dukic, C. Ahlstrom, C. Patten, C. Kettwich, K. Kircher, Effects of Electronic Billboards on Driver Distraction, Traffic Inj. Prev. **14** pp. 469–476 (2013). doi:10.1080/15389588.2012.731546
3. K. Domke, K. Wandachowicz, M. Zalesińska, S. Mroczkowska, P. Skrzypczak, Ocena zagrożeń występujących w ruchu drogowym powodowana przez wielkopowierzchniowe reklamy z diodami świecącymi, work no. 42-17-223-42050, grant no. RoM. III/3420-50/10 (Poznan, 2010)
4. Act of 21 March 1985 on Public Roads, OJ 2007 No. 19 Pos. 115 (Poland, 1985)
5. Guideline: Roadside Advertising Guide, Edition 1.2 (Departament of Transport and Main Roads, Australia, 2013)
6. S. Speirs, A. Winmill, T. Kazi, The Impact of Roadside Advertising on Driver Distraction: Final Report (WSP Development and Transportation, Basingstoke, 2008)
7. M. Zalesińska, K. Wandachowicz, The research of outdoor billbords with light emitting diode using luminance measuring camera, Pozn. Univ. Technol. Acad. Journals. Electr. Eng. **69** pp. 275–282 (2012)
8. Roadside advertising manual, Document No: AD-R-03 First Edition (Deparment of Transport United Arab Emirates, Abu Dhabi, 2013)
9. I. Lewin, Digital Billboard Recommendations and Comparisons to Conventional Billboards (Lighting Sciences Inc., Arizona, 2008)
10. Effects of roadside advertisements on road safety, Finnra Internal Reports 25/2004 (Helsinki, 2004)

11. P.R. Boyce, Lighting for driving: roads, vehicles, signs, and signals (Taylor & Francis Group, USA, 2009)
12. D. Beijer, A. Smiley, M. Eizenman, Observed Driver Glance Behavior at Roadside Advertising Signs, Transp. Res. Rec. J. Transp. Res. Board. **1899** pp. 96–103 (2004). doi:10.3141/1899-13
13. Act of 24 April 2015 amending certain acts on strengthening the tools for landscape protection, OJ 2015 Pos. 774 (Poland, 2015)

Evaluation of Intelligent Transport Systems impact on school transport safety

Dagmara Jankowska-Karpa[1,*], and *Justyna* Wacowska-Ślęzak[1]

[1]Motor Transport Institute, Road Safety Centre, Warsaw, Poland

Abstract. The integrated system of safe transport of children to school using Intelligent Transport Systems was developed and implemented in four locations across Europe under the Safeway2School (SW2S) project, funded by the EU. The SW2S system evaluation included speed measurements and an eye-tracking experiment carried out among drivers who used the school bus route, where selected elements of the system were tested. The subject of the evaluation were the following system elements: pedestrian safety system at the bus stop (Intelligent Bus Stop) and tags for children, Driver Support System, applications for parents' and students' mobile phones, bus stop inventory tool and data server. A new sign designed for buses and bus stops to inform about child transportation/children waiting at the bus stop was added to the system. Training schemes for system users were also provided. The article presents evaluation results of the impact of selected elements of the SW2S system on school transport safety in Poland.

1 Introduction

The most critical part of the children's route to school and back is the way to/from the bus stop and the area around the stop. It is therefore essential to develop a system that would assist the bus driver, providing them with information about the route, the location of bus stops and the children who should get on/off at the stop. In addition, it is important to improve the visibility of school buses and raise road user awareness by using a uniform sign of school buses in the European Union, which would result in other vehicles reducing their speed or stopping completely when children are boarding or leaving a school bus nearby. The use of Intelligent Transport Systems (ITS) in school transport may also improve the safety of children in traffic. All these elements have been taken into account during the implementation of the Safeway2School – SW2S (Integrated system for safe transport of children to school) European project.

2 European Project Safeway2School – Integrated System for Safe Transport of Children to School

The SW2S project was implemented under the 7th Framework Programme of the European Commission. In Poland, it was led by the Road Safety Centre of the Motor Transport Institute. It was coordinated by the Swedish VTI Institute and implemented by fourteen partners from seven European countries.

* Corresponding author: dagmara.jankowska@its.waw.pl

Under the project, a route planning system, rerouting system and warning system were developed for vehicles in the vicinity of school bus stops. It was also necessary to create a network of intelligent bus stops - IBS (enabling communication between child, stop and bus) and to equip the children participating in pilot studies with special transmitters (tags) that activated bus stop lights.

The target group of the project included primarily school bus drivers, students aged 6 to 16 who use school transport and their families as well as infrastructure managers, vehicle manufacturers, other road users and public administration representatives.

The purpose of the pilot undertaken within the framework of the project was to develop guidelines and recommendations as well as training courses on safe school transport.

Four countries were selected for the pilot implementation of individual Safeway2School system components: Sweden, Austria, Italy and Poland. After the completion of the project, the utility and effectiveness of the technologies employed were evaluated and their acceptance among system users was studied [1].

Figure 1 shows a schematic diagram of the entire Safeway2School system whose components were tested in four pilot locations.

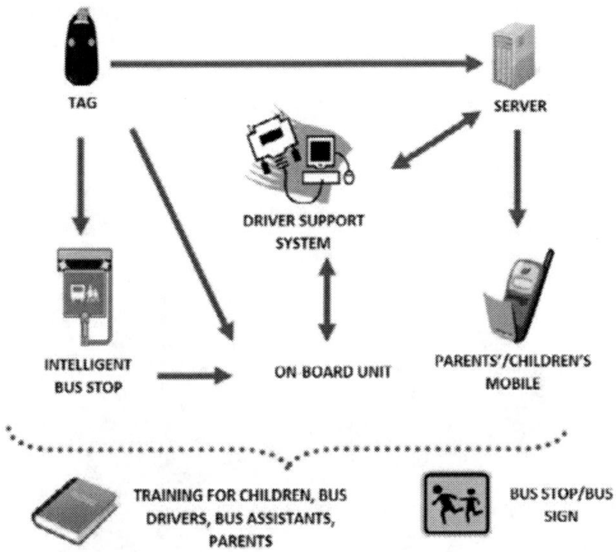

Fig. 1. Diagram of the SW2S integrated system for safe transport of children to school [2].

The Safeway2School system comprises the following components [2]:

- The system for protecting pedestrians at the bus stop and around it – the intelligent bus stop (IBS) and the tag carried by the child. The IBS warning module is powered by solar energy. When the child approaches the stop, his or her tag (using a radio signal) activates the module, which starts flashing its lights.
- The system for assisting the driver interfaces with the bus's on-board computer, using a special application containing information e.g. on the bus route (the bus can be rerouted if unforeseen traffic incidents occur), its speed, stops on the route and the children getting on/off the bus.
- The application installed on the parent's mobile phone, which makes it possible to notify the parent by text message that the child has boarded the bus and has reached his or her school. It is also possible to alert the driver that the child will not be present at the bus stop on a given day.

- The application installed on the child's mobile phone enables the phone to operate as a transmitter, notifying parents that the child has boarded the bus and has reached school/home.
- Bus stop inventory tool – after inputting data related to e.g. traffic around the bus stop, the speeds of vehicles in its vicinity, the number of people waiting, etc., the safety level of the bus stop can be determined.
- The server which collects all data from the entire system.

The SW2S system is supplemented by specially developed training courses for children, their parents and school bus drivers and also by a special sign developed under the project, which was placed on the school bus and on the stops along its route.

3 Pilot Implementation of the SW2S System in Poland

The purpose of the pilot implemented in Poland within the framework of the SW2S project was to study the impact of the new road sign, new school bus sign and IBSs on the behaviour of other road users. The studies focused on drivers of the vehicles that passed the new road sign and school buses with new sign and IBSs. The pilot was carried out in Nowy Dwór Mazowiecki. In the selected primary school, ca. 170 children (out of ca. 500 students attending the school) use the school bus on a daily basis. Among them, 47 were selected for the pilot (in two age groups: aged 7–9 and 10–12) together with their parents and a school bus driver. As part of the pilot, the following SW2S System components were tested in Poland [2]:
- IBSs – 3 stops on the school bus route;
- children's tags – 47 children;
- new bus stop sign – placed on all stops along the school bus route;
- new school bus sign – on the front and rear of the bus.

Fig. 2. Elements of the SW2S System tested in Poland [2].

The SW2S system was studied in three stages:
- pre-test: before system implementation (vehicle speeds in the vicinity of selected bus stops);
- pre-pilot: installation and testing of system components;
- pilot: system implementation and evaluation.

4 Impact Assessment of SW2S System

The following components of the SW2S system were evaluated in Poland:
- impact of new school bus sign:
  - parents, children, school bus driver – sense of safety assessment – questionnaires;
  - drivers of other vehicles:
    - attention and perception assessment – questionnaires;
    - tests using eye tracking equipment;
- impact of IBSs on traffic:
  - the speed of vehicles passing IBSs (before and after the introduction of the system);
  - parents, children, school bus driver – sense of safety assessment – questionnaires;
- training:
  - parents, children, school bus driver – sense of safety assessment – questionnaires.

In this article, only the procedure and results of the experiment with the use of eye tracking equipment is presented.

5 Eyetracking Experiment

The purpose of the study involving the eye tracking equipment was to assess the effectiveness of IBSs and of new school bus sign in terms of their visibility to motorists driving past the IBSs and past the bus with the new sign [1]. Sixteen subjects participated in the study. The subjects were drivers with category B driving licences. Most of them did not know the road on which they drove during the study. The study involved two buses: one with a new SW2S sign and one with the sign mandated in Poland. The study also covered a bus stop with the IBS module and a stop without this module. Both stops had the new SW2S sign. Each subject encountered the same scenario on the route and the sequence of events was the same in each case.

5.1 Study Procedure

The route along which the study was conducted started at stop No. 1 (close to the school) and terminated at stop No. 8 – Wymysły. Travel time amounted to ca. 20 minutes in both directions. A bus with the old sign was stationary at stop No. 7a (waiting for the experimental car) and activated the IBS. The experimental car passed a bus with the new sign, which moved in the opposite direction, close to stop No. 4a. Subsequently, it passed the stationary bus at stop No. 7a (IBS lights were active), passed the last stop (No. 8) and then turned back. When it was passing stop No. 4, the IBS was active and a passenger (pedestrian) was waiting at the stop. Then it passed stop No. 2, which did not include the IBS module but a person was waiting there as well.

 DOI: 10.1051/matecconf/201712203003

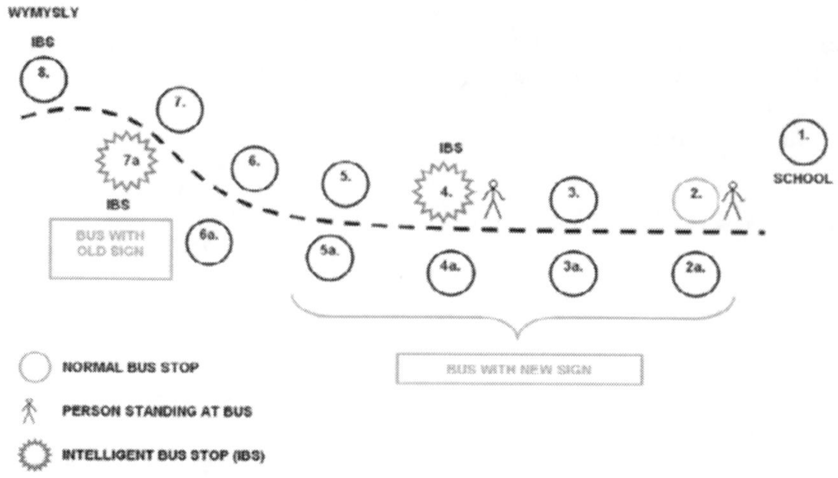

Fig. 3. Scheme of eye tracking experiment route [2].

The study using the eye tracking equipment covered the following situations:
B1: A bus with the new sign follows its normal route from stop No. 1 to stop No. 8 and back, stopping at bus stops. The experimental car passes it in the area of stop No. 4a.
B2: A bus with the old sign is stationary at a stop with the IBS module (No. 7a).
H1: Children are waiting for the bus at IBS stop – the warning module is active (4).
H2: Children are waiting at a regular stop (2).

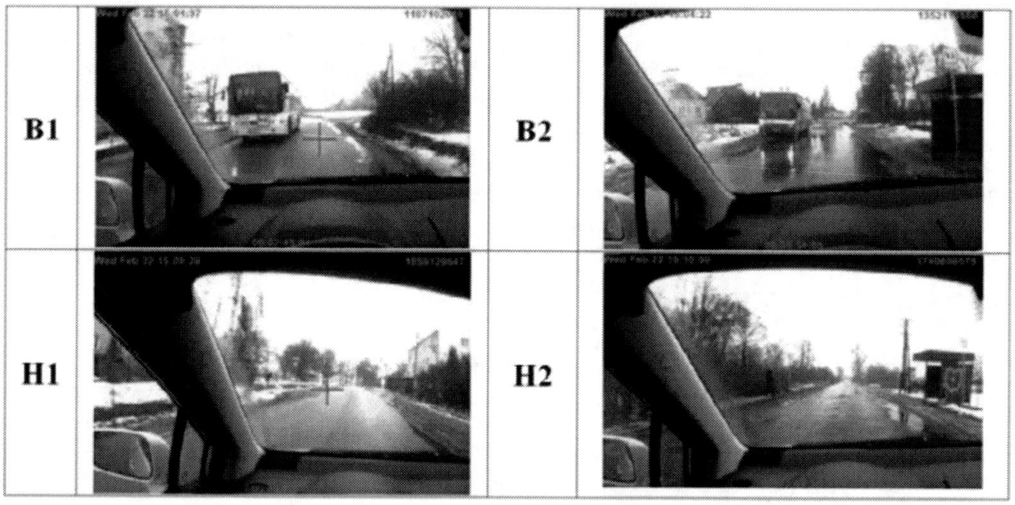

Fig. 4. Situations in the eyetracking experiment [2].

5.2 Hypotheses

The hypotheses related to the study were formulated as follows [4]:
- The time during which subjects (experimental car drivers) look at the target (the bus and new signage on the bus) is longer for B1 than for B2 as the new sign is more visible than the one presently mandated. Also, the first look at the new sign occurs earlier than at the old sign.

- The first look in situation H1 occurs earlier than in situation H2 because the stop with IBS and flashing lights is more visible than a regular bus stop.
- The time during which subjects look at children at IBS (H1) is longer than while looking at children at a regular stop (H2) because IBS is more visible than a regular stop.

5.3 Study Results

The numbers in Tab. 1 represent the total time from the first look until the target is passed and the time during which the driver's gaze is fixed on the target in the four analysed situations. Each column in Table 1 represents the mean value for the entire study.

Table 1. Total time from the first look until the target is passed and Time during which driver's gaze is fixed on target (ms) in 4 analysed situations [2].

	Total time from the first look until the target is passed (ms)	Time during which driver's gaze is fixed on target (ms)
B1	17528	3161
B2	18080	4513
H1	24320	2503
H2	19615	2020

The total time from the first look until the target is passed is the interval between the time when the subject looks at the target (bus/signage on the bus/children) and the time when the target is passed. This indicator is used because the distance in metres cannot be calculated. The time during which the driver's gaze is fixed is the total time during which the driver looks at the target (bus/signage on the bus/children).

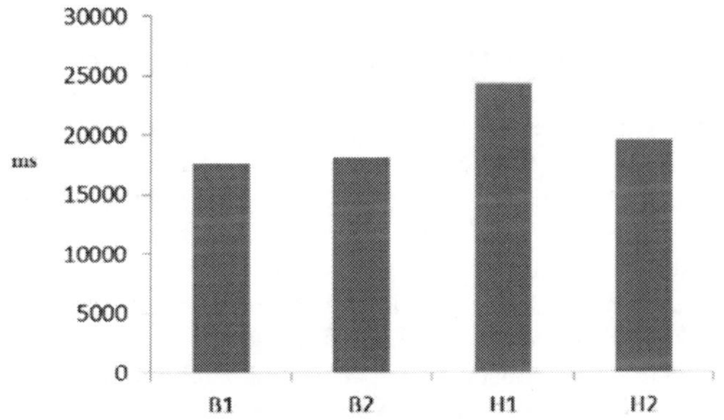

Fig. 5. Total time from first look until the target is passed (ms) [2].

The time of the first look at the target is shorter for situation B1 than for B2. However, in this case the comparison is subject to an error because in situation B1 the bus was moving and, moreover, moving in the direction opposite to the experimental car driver, which reduced the time available. Therefore it is appropriate to assume that the driver spent more time looking at B1 than at B2, which confirms the hypothesis previously adopted.

For situation H1, the total time from the first look until the target is passed is longer than the time for situation H2, which also confirms the hypothesis previously adopted, and the difference amounts to 4,705 ms. The drivers looked at the bus stop longer when IBS lights were active compared to a stop without the IBS module.

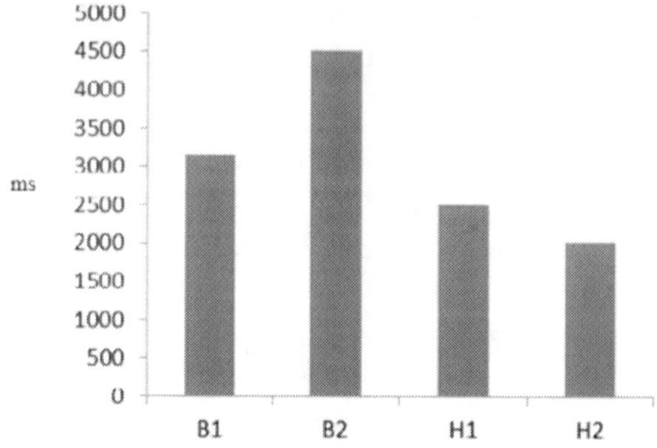

Fig. 6. Time when gaze is fixed on target (ms) [2].

The time during which the driver's gaze was fixed was shorter for situation B1 than for B2, but – as mentioned above – this was the case because in situation B1, the bus was in motion and moving toward the driver. Comparing situations H1 and H2, the time during which the driver's gaze is fixed is longer for H1 (by ca. 483 ms), which means that the driver paid more attention to the target (child + IBS) than to the same child waiting at the bus stop without active IBS lights.

Studies on drivers using the eye tracking equipment have successfully confirmed the hypotheses adopted [2].

6 Conclusions

The results of the study conducted using the eye tracking equipment indicate that the intelligent bus stop (IBS) attracts more attention from drivers than a regular one. Active bus stop lights cause drivers to pay more attention to what is happening on the road and they notice such stops earlier than regular bus stops without the IBS module.

The results of the study concerning the time when the sign on the bus is noticed and processed are difficult to interpret since the bus with the new sign was in motion. However, slight differences in the total time that elapsed from the first look until the target was passed suggest that drivers noticed the new sign on the bus earlier than the old sign, which also confirms the hypotheses adopted. Additionally, each experimental car driver was asked to complete a short questionnaire after driving the route. The questions concerned, among others, the subjects' knowledge of the speed limits applicable on the route (most drivers knew them, with slight inaccuracies). Most drivers (80%) paid attention to the new sign on the bus and also to flashing IBS lights (75%). Most of them also declared that they always slow down when they see people waiting at a bus stop or a bus approaching a bus stop. Two persons declared that they had reduced their speed considerably when seeing flashing lights at the bus stop (IBS).

It should be emphasised that the expected speed reduction of other vehicles on the route was also achieved during IBS activity.

Elements of the Safeway2School system have been evaluated favourably in the studies, which is of considerable importance for ensuring child safety on the route to school and improves road safety.

References

1. D. Jankowska, Integrated system for safe transportation of children to school -with the use of Intelligent Transport Systems (ITS), J. KONES Powertrain Transp. **15** (2008)
2. D. Jankowska, J. Wacowska-Ślęzak, Report from pilot implementation of SAFEWAY2SCHOOL system in Poland (Warsaw, 2012)
3. A. Bulling, J.A. Ward, H. Gellersen, G. Tröster, Eye movement analysis for activity recognition, in: Proc. 11th Int. Conf. Ubiquitous Comput. - Ubicomp '09, pp. 41–50 (ACM Press, New York, USA, 2009). doi:10.1145/1620545.1620552
4. A. Bojko, A. Stephenson, It's All in the Eye of the User: How Eye Tracking Can Help Answer Usability Questions, User Exp. **4** (2005)

Impact of ITS services on the safety and efficiency of road traffic

Tomasz Kamiński [1,*], *Mirosław* Siergiejczyk[2], *Jacek* Oskarbski[3], and *Przemysław* Filipek[4]

[1]Motor Transport Institute, Road Traffic Safety Centre, Warsaw, Poland
[2]Warsaw University of Technology, Faculty of Transport, Warsaw, Poland
[3]Gdansk University of Technology, Faculty of Civil and Environmental Engineering, Gdansk, Poland
[4]Lublin University of Technology, Faculty of Mechanical Engineering, Lublin, Poland

Abstract. This paper describes the project entitled: "Impact of the use of Intelligent Transport Systems on the level of road safety" (agreement no. DZP/RID-I-41/7/NCBR/2016), implemented as part of the Road Innovations Development (RID) programme, funded by the National Centre for Research and Development and the General Directorate for National Roads and Motorways (GDDKiA). The project is run by a consortium comprising the Motor Transport Institute, University of Gdańsk, Warsaw University of Technology, Military University of Technology and Research Institute of Roads and Bridges. The impact of ITS services and accompanying modules on road safety may be considered in functional, logical or physical terms. The aim of the project is to understand how ITS services change road safety.

1 Introduction

The project entitled: "Impact of the use of Intelligent Transport Systems on the level of road safety" (agreement no. DZP/RID-I-41/7/NCBR/2016) is implemented as part of the Road Innovations Development (RID) programme, funded by the National Centre for Research and Development and the General Directorate for National Roads and Motorways (GDDKiA). The project is run by a consortium which includes the Motor Transport Institute, University of Gdańsk, Warsaw University of Technology, Military University of Technology and Research Institute of Roads and Bridges. The impact of ITS services and accompanying modules on road safety may be considered in functional, logical or physical terms.

The functional aspect is investigated based on the influence of services and groups of services on road user behaviour and the homogenization of the traffic stream. Both contribute to a change in the level of road safety, characterized by a reduction in road traffic hazard. Many years of research conducted in the US, Japan and Europe have shown that ITS-based services and services designed to inform travellers using variable message signs, boards, or traffic lights (at intersections and entries to motorways or express roads - in the case of ramp metering) have helped to reduce accidents by approx. 80% using advanced traffic management systems, by 25-50% using ramp metering and 30-40% using traffic control via variable message signs [1–4]. Automated traffic surveillance (speed limit enforcement) contributes to a reduction in accidents by 20-80%.

* Corresponding author: tomasz.kaminski@its.waw.pl

Improving traffic safety and efficiency is logically associated with improving cooperation between road managers and emergency services and with implementing solutions designed to reduce rescue operation time in order to reduce accident severity, minimize the period of exposure to the risk of secondary events and minimize the loss of time of the travellers. According to the principle of the "Golden hour" the life of 20-40% of seriously injured victims can be saved if they receive hospital treatment within 60 minutes of the event. The probability of survival is greater when first aid is provided quickly at the scene prior to being transported to hospital (during the "Golden ten minutes").

Based on the results of European research, it is estimated that transport telematics can reduce the response and intervention time of rescue services by up to 30% (US studies show a 20-40% shortening of the rescue operation time), and the use of emergency calls, automatically generated by in-vehicle systems increases the probability of survival of accident victims by 15% [5,6]. Linking traffic management with a rescue operation is very effective. The traffic events management systems (including automatic detection, verification of events, guiding the emergency services and restoring normal traffic conditions) contribute to shortening detection time by up to 65%, shortening the time needed for emergency services to arrive by about 45% and reducing secondary accidents by 7-50% [4].

The level of road safety is also affected by the use of ITS services in private vehicles (collision avoidance support systems, monitoring the condition of the driver, etc.) and emergency vehicles (e.g. automatic vehicle location, navigation to the scene of the incident, traffic lights priority). Indirectly, the modules associated with the physical layer (reliability of the equipment and IT systems) are important for the overall security of ITS systems. The above effects of implementing ITS services have been elaborated based on statistical surveys carried out in country-specific conditions. Due to the diversity of the evaluated solutions, an approximate assessment of their impact on road safety was developed.

The project "Impact of the use of Intelligent Transport Systems on the level of road safety" was initiated due to a lack of assessment methodology. The project plans to use advanced driving simulators and simulation software Vissim/Saturn/Visum to obtain vehicle motion parameters such as velocity, acceleration, braking power, velocity profile, etc. as the reaction of drivers on ITS services. These parameters will be used in safety models for evaluating the level of road safety.

Another project task is to use the multi-criteria Analytic Hierarchy Process (AHP) method to understand how ITS services change road safety and traffic efficiency defined as a stream of vehicles driving in a specific period of time.

Both project tasks are currently under way and will be described in papers to come. This will help to define further research directions.

2 Project assumptions

The objective of the project described in the article is to develop tools designed to evaluate the impact of the expected Intelligent Transport Systems solutions on road safety, particularly in the context of the implementation of the National Traffic Management System. For the purpose of the project it is assumed that the level of road safety (as related to ITS services) will be mainly affected by the following factors, depending on the scope and type of implemented ITS solutions:

- features (factors) related to the road and the traffic organization (among others road class, geometric solutions and traffic layout solutions),
- traffic features (among others speed, density, intensity, utilization of throughput),
- road illumination,
- weather conditions and the state of the road surface,
- reliability of the physical layer (in terms of hardware and ICT),

- features (factors) associated with the driver behaviour (i.e. driving speed and failure to adapt it to the road and traffic conditions.

It is understood that traffic efficiency can be characterized by traffic flow volumes. The impact of ITS services on the level of road safety will be determined using advanced driving simulators and Visum/Saturn/VisSim simulation software and traffic efficiency will be determined using simulation software which depends on the specifics of the adopted assessment method.

Driving simulators are used to evaluate the road cross-section and road sections. The simulation software helps to assess the road corridor (road network area).

The project will be implemented mainly in relation to the area covered by the National Traffic Management System (KSZR), which will be implemented in 17 modules:

- Module 1 – "Transmission of information and instructions for drivers",
- Module 2 – "Output to control the speed and traffic lanes",
- Module 3 – "Collecting data about the vehicles",
- Module 4 – "Event detection from the available resource data",
- Module 5 – "Collecting data on the journey",
- Module 6 – "Motorway Alarm Telephony + CB",
- Module 7 – "Video data",
- Module 8 – "Transmission of how busy MSA/Car Parks are",
- Module 9 – "Obtaining information on how busy MSA/Car Parks are",
- Module 10 – "Managing road lighting",
- Module 11 – "Dosing entry (Ramp Metering)",
- Module 12 – "Traffic lights on the roads",
- Module 13 – "Collecting weather data and the road surface condition data",
- Module 14 – "Noise measurement",
- Module 15 – "Air pollution measurement",
- Module 16 – "Transmission of information for motorists about the tunnel",
- Module 17 – "Collecting traffic data from the national road network".

These modules will have a direct impact (e.g. by displaying information on traffic conditions, or a dynamic introduction of speed limits - variable contents signs) or indirect impact (e.g. by obtaining information about the weather conditions and the state of the road surface) on road safety. The implementation of Intelligent Transport Systems within the roads managed by the GDDKiA is currently under way. The implementation work must be preceded by research and development that will help select optimal solutions (meeting the criteria developed) to be justified in terms of implementation costs. In this way it will be possible to achieve the maximum cost benefits.

According to the strategic objectives of the National Traffic Management System:

- ITS systems should address the needs of their users;
- requirements for the systems should take into account the real needs, rather than market trends, or the current capabilities of suppliers;
- data exchange between the public sector significantly increases the efficiency of ITS investments;
- experiences in other countries and the technology market are rapidly changing, hence the need for cooperation with the ITS sector, scientific community and other units of the road administration needs to be addressed.

There is no scientific proof of how individual ITS devices influence the recipients, i.e. road users or the proper way to deploy these devices within road infrastructure. This gives us some freedom to choose functional, organizational, hardware and telecommunications solutions of these systems, which are characterized by a certain dissimilarity in how the information is provided to the end users (road users). They also differ in terms of the message content.

There are no guidelines or recommendations in this respect at the European and national level. This justifies the need for the project to analyse the impact of ITS systems on road safety. It would be wrong to assume that road statistics alone with injury and serious injury numbers are sufficient for the purposes of such an evaluation. ITS systems can be highly complex and feature mutual interactions, interactions with the rest of the road infrastructure, emergency systems, in-vehicle on-board systems and ITS solutions implemented in cities.

To examine the impact of a single ITS system or service, based on the statistics of time period-based road accidents, would require long-term studies, assuming that other technical and traffic conditions on the road have not changed significantly. Even 2-3 years of road safety research on a section with an ITS system would not give a clear answer to the question about the impact on road safety or traffic efficiency. Let us take the example of a road section, on which two accidents occurred over two years, in which two people died. In year three a bus accident happens leaving 17 people dead. The analysis of statistical data might suggest that the danger on the road measured by the number of accidents increased 17 times in the last year, compared to the previous 2-year period. It is also possible to evaluate safety based on specific events, for example, traffic conflicts. In most cases, however, this method would not be reliable.

3 Evaluation of the impact of ITS services on road safety

As part of the above work, the analysis will take into account ITS solutions which affect traffic flow and safety significantly. The comparative studies of the impact made by ITS services will be made along the selected road stretches in the course of data analysis from a real ITS system.

The comparison will be based on vehicles speeds along the stretch of the road in a situation when the systems would not display any information for the drivers and a situation when such information would be displayed. The influence of the displayed content on the driver behaviour will be evaluated.

In a series of experiments with advanced driving simulators the listing of Variable Message Signs will be made along the section of the road on which the actual ITS system is functioning (the section dealt with in the first phase of work). This will be the basis for developing research scenarios. Next, a simulation will be made involving drivers, during which the data will be recorded to assess the impact of ITS systems on the drivers. The data will then be analysed for quantity (vehicle speed, vehicle speed change, etc.) and quality (lane change, compliance or failure to comply with the recommendations and commands of the road administrator, etc.) in order to assess the impact of ITS services on road safety.

Finally, an experiment will be conducted to assess the impact of ITS services on traffic flow and road safety, using simulation Visum/Saturn/VisSim software. Simulations will be performed both on the micro scale (road section, a single intersection) and area. The models which will be used for the simulation will be calibrated using data from the real-life advanced traffic management system. Various system configurations providing various ITS services will be studied. It is expected that the models will be verified using data from the real ITS system.

Despite the well-known limitations and disadvantages of the proposed simulation methods, it is estimated that the evaluation of the impact of ITS services on road safety and traffic efficiency will be a better option than a similar evaluation using incomplete and insufficient volume of data from the actual ITS systems.

It will not be possible in every case to evaluate the impact of ITS services on Poland's road safety using data from foreign ITS systems. This is because of a slightly different behaviour of drivers and some other legal regulations in the field of road traffic. Based on the previous experiments, indicators will be developed in order to assess the impact of ITS services on the effectiveness of traffic and road safety as well as the guidelines and recommendations for their practical application and multi-criteria method to assess impact of ITS services on road safety and traffic effectiveness.

Currently, the impact of ITS services on road safety is assessed primarily based on the individual assessment of experts, statistical surveys or the few road safety models which still require more work. There is no structured or uniform assessment method that would provide an opportunity to compare the effect of ITS services and their different configurations on road safety and traffic efficiency.

The advantage of the proposed solution is the use of a holistic approach to the abovementioned problem which involves a combination of:

- research based on data and information from real advanced traffic management system incorporating a variety of ITS services (e.g. passing the information over to the drivers, traffic control, etc.);
- studies using advanced driving simulators, with experienced road transport psychologists to interpret test results;
- tests using Visum/Saturn/VisSim vehicle movement simulation software.

The unique combination of three different approaches will enable a comprehensive analysis of the impact of ITS services on road safety and traffic efficiency. It is envisaged as a two-stage evaluation. The first stage will be carried out by an assessor, not necessarily an expert in the field of road safety and ITS services. They will use the multi-criteria evaluation method, developed under this project, evaluation criteria and guidelines for the evaluation and selection of ITS services.

The second stage of evaluation will be carried out in exceptional cases where a clear-cut evaluation of the impact of ITS solutions is not possible. This may be due to the complexity of the system, numerous interactions between ITS services and the rest of the road infrastructure and the road surroundings (including local conditions). Then, an expert evaluation will be carried out using advanced driving simulators and the traffic simulation software.

4 AHP method

The multi-criteria AHP (Analytic Hierarchy Process) method was proposed by Thomas L. Saati [7] and applied in many areas. It allows a hierarchical analysis of the decision problem to choose one of the variants or objectively compare the proposed solutions. Variants may be technical objects with defined parameters as well as certain conditions associated with, for example, safety, risk or quality. The AHP method provides a multi-criteria approach that takes into account the evaluator's preferences. Unlike objective measurements of the magnitude of a solution it makes it possible to take into account the preferences of the evaluator, treating them as a natural phenomenon for the choices made by a human. This method allows for taking into account not only the quantitative parameters characterizing the given phenomenon, but also the qualitative characteristics associated with the desired properties of the object.

The hierarchical structure of the decision-making process consists of levels of purpose, criteria, sub-criteria (optional) and variants. The decision maker's judgment is determined by the relative importance ratings determined by comparing pairs of all objects from the hierarchy level. Ratings are expressed in numerical values, most often on a scale of 1 to 9.

The use of the AHP method includes:

- Building a model of a hierarchical object together with determining the importance of factors (criteria).
- Evaluating by pair comparison.
- Designation of general and personalized preferences.

- Classification of decision options in terms of the degree of achievement of the desired goal.

In the case of the RID 4D project, the AHP method will be used to combine an evaluation of the solutions, individual ITS services obtained, among others by simulations using advanced driving simulators, evaluations from the Vissim/Staturn/Visum simulation software and technical parameters (including telecommunications). Solutions will be applied taking into account the cost criterion of implementation. The purpose of the evaluation will be to provide the most favourable ITS services or a combination of services in a variety of road-motion conditions.

5 Implementation of the project results

The project results will be implemented as officially announced by the General Directorate for National Roads and Motorways (GDDKiA). As a final project result, there are plans to organize training courses for a group of 30 GDDKiA employees. The training will cover the practical application of the multi-criteria method developed in the project and assessment of ITS services impact on road safety and traffic efficiency. The trainees will be able to apply this method in conjunction with the indicators and guidelines for evaluation. The project results will also apply to the evaluation of ITS services on the roads not managed by the GDDKiA.

The project consortium also plans to organize a scientific conference to present the project results.

There will be a monograph on evaluating ITS systems' impact on road safety and traffic efficiency in a comprehensive and exhaustive manner. The monograph will be developed based on the results of the project after its completion.

After completion of the project, an offer will be prepared presenting the effects of the project to stakeholders. The stakeholders will primarily include managers of roads run by local governments and commercial companies offering ITS services provided occasionally under extensive nationwide systems.

An additional effect of the project will be the launch of comprehensive services in the field of assessing the impact of ITS services on road safety and traffic efficiency at the Motor Transport Institute and the Gdansk University of Technology. The service will incorporate the effects of the project, combined with advanced driving simulators and Visum/Saturn/VisSim software. The evaluation models will be developed based on the experience gained during the analysis as well as the data relating to the actual ITS services operating as part of the advanced traffic management system.

Acknowledgements

The article is part of the project RID-4D (41) "The impact of the use of Intelligent Transport Systems services on the level of road safety" (OT4-4D/ITS-PW-PGIBDiM-WAT), funded by the National Centre for Research and Development (NCBR) and the General Directorate for National Roads and Motorways – GDDKiA (contract no. DZP/RID-I-41/7/NCBR/2016 of 26 February 2016).

References

1. ITS Handbook 2000 – 2001. Supervised by the Ministry of Construction (Tokio, Japan, 2001)
2. ITS Handbook 2000. Recommendations from the World Road Association (PIERC) (Artech House, Boston, 1999)
3. J. Bunch, C. Burnier, E. Greer, G. Hatcher, A. Jacobi, F. Kabir, C. Lowrance, M.

Mercer, K. Wochinger, Intelligent Transportation Systems Benefits, Costs, Deployment and Lessons Learned Desk Reference 2011 UPDATE. RITA Intelligent Transportation Systems Joint Program Ofiice. Final Report – September 2011. FHWA-JPO-11-140 (2011)

4. A.T. Proper, Intelligent Transportation System Benefits: 2003 Update (U.S. Department of Transportation, Washington D.C., 2002)
5. ERTICO, Expected Benefits of ITS, in: 4th World Congr. ITS, (Berlin, Germany, 1997)
6. M. McDonald, H. Keller, J. Klijnhout, V. Mauro, R. Hall, A. Spence, C. Hecht, O. Fakler, Intelligent Transport Systems in Europe. Opportunities for Future Research, World Sci. (2006)
7. T.L. Saaty, L.G. Vargas, Decision Making with the Analytic Network Process. Economic, Political, Social, and Technological Applications with Benefits, Opportunities, Costs and Risks, Springer Sci. Media. (2006)
8. D. Pawlus, A. Tarnowski, R. Lewkowicz, T. Kamiński, M. Kruszewski, Innowacyjny zestaw metod i narzędzi do analizy infrastruktury drogowej w aspekcie poprawy bezpieczeństwa, Logistyka. **6** (2014)
9. M. Niezgoda, M. Kruszewski, T. Kamiński, A. Tarnowski, Operationalization of DBQ measures in the driving simulation environment, in: Int. Conf. Road Saf. Simul., (Rome, 2016)
10. E. Łazowska, T. Kamiński, M. Niezgoda, M. Kruszewski, R. Grzeszczyk, The concept of automatic road safety evaluation, Combust. Engines. **52** pp. 728–734 (2013)

The safety issue of roadside advertising – comparison of Polish and Abu Dhabi regulations

Tomasz Mackun[1,*], and *Joanna* Żukowska[1]

[1]Gdansk University of Technology, Faculty of Civil and Environmental Engineering, Gdansk, Poland

Abstract. In Poland a large number of advertisements are located by the roadside. These ads do not support road traffic management and unlike the road marking system are not subject to any regulations. The advertiser's goal is to communicate a message to as many recipients as possible. Drivers with different individual abilities, such as attention focusing, eye accommodation, speed of information processing, can be distracted, blinded or confused by the content and form of the advertising. There are elements of the road network, such as intersections, pedestrian crossings, road junctions etc. where the driver must assess the situation on the road, predict the behaviour of other users, make decisions and finally complete a manoeuvre. It all happens in a limited span of time when actions should be taken calmly with full attention. It is obvious that the attention of drivers, especially in those zones, should be focused on the task of driving. In this article, the authors present a perspective on selected national laws, and also quote Abu Dhabi's advertising placement manual [1] as a good example of how to manage roadside advertising.

1 Introduction

The advertiser's goal is to communicate a message or idea to the largest possible audience at minimal cost. Promoting products and ideas has been known for centuries. Competition generates brand building, requiring information on products, pricing and address data. One form of advertising is external advertising. Outdoor advertising is best used in places where the largest possible audience can see it, hence a natural location is by roads and streets.

In Poland after 1989 and the collapse of communism a free market began to emerge and this resulted in marketing activities, including the promotion of products. For nearly 30 years, there has been a continuous, practically unlimited practice of placing advertising at the roadside. The greater the traffic volume, the more willing the advertisers are to use this location for their advertising. It is standard practice to measure the traffic volume on multiple roads in order to situate advertising on the road with the highest possible traffic volume.

Public perception of outdoor advertising is becoming more negative due to the increasing density of advertising in public spaces and the lack of regulation that determines the location, form, content and aesthetics. There are many social movements that try to fight against advertising, especially those that represent the belief that advertisements deface the environment and are located illegally.

* Corresponding author: tomasz.mackun@pg.gda.pl

Unfortunately, for years in Poland, advertising has not been analysed as potentially distracting for drivers. Distraction may deteriorate driving performance. Therefore, it is thought that advertising can deteriorate road safety.

The answer to public demand for regulation of outdoor advertising is the so-called Landscape Act, i.e. the Act of 24 April 2015 amending certain acts. It strengthens landscape protection tools [2]. This law allows municipalities, under local law, to adopt resolutions which govern the conditions of advert placement and the collection of appropriate fees. Unfortunately, the Act [2] did not take into account the need to evaluate roadside advertising as objects that could distract drivers and pose a greater risk on the road.

Many countries prohibit roadside advertising, and others have implemented appropriate regulations. The examples are the Australian instruction guide [3], the Abu Dhabi manual [1], and the instructions for national roads in South Africa [4]. These instructions introduce rules that relate to road traffic safety. Their use is intended to minimise distraction, confusion or dazzling of drivers by advertisements.

2 The Polish perspective

In 1984 Poland ratified the European Agreement on Main International Traffic Arteries (AGR) [5], signed in Geneva on 15 November 1975, which prohibits the placement of advertisements on international roads. This regulation, however, is ineffective outside the boundaries of the carriageway. In Poland, regulations rarely refer directly to roadside advertising. More often, if an advertisement is considered a permanent structure on the ground, it is subject to building regulations. Unfortunately, these regulations rarely apply to traffic safety. While the Public Roads Act of 21 March 1985 [6] introduces regulations, it must be regarded as insufficient. This law regulates only the roadway and a small area in its vicinity. For drivers the boundary of the roadway is an invisible line. Existing regulations specifically regulate advertisements on the carriageway, however the placement of adverts immediately adjacent to the roadway is not regulated at all. The Act [6] forbids the placement of adverts on the carriageway outside built-up areas but allows for locating them on the carriageway in built-up areas. The rule prohibits the placement of advertisements that imitate traffic signs, markings and traffic safety devices. Unfortunately, the rules do not mention motorist distraction. The amendment to the Act on Public Roads [6], in connection with the Landscape Act [2], introduces regulation for off-carriageway advertising at a distance in accordance with Table 1, in terms of light emission, display of animated images, visual effects, gaps between successively displayed information and also minimum display of visual information presentation of 10 seconds. Unfortunately, the distance shown in Table 1 is insufficient for most roadside advertising to be affected. "The Regulation of the Minister of Infrastructure and Development [7] of 20 October 2015 on the technical conditions to be met by level crossings and their location" prohibits the placement of advertising within the visibility triangles and on a section of 20 m from each side of the level crossing, as well as on railway viaducts within the area of the roadway.

The Act [8] on the Road Traffic Law prohibits "the placing of objects that emit or reflect light within the road boundaries in a way that may blind road users or mislead them." The provision of this law is appropriate but very difficult to enforce. There is also no provision for banning devices that may distract or mislead drivers.

The Convention on Road Signs and Markings [9], drawn up in Vienna on 8 November 1968 which the Polish Government ratified specifies that "[...] parties take measures to prohibit placement of boards, posters, signs or devices that can be confused with signs or other traffic management devices, or may reduce their visibility or efficacy, or dazzle the road users or distract them in ways that threaten road safety [...]". This is the only document that mentions driver distraction but secondary legislation within this scope was not implemented.

Table 1. Distance from the road boundaries where advertising cannot be located.

No.	Type of road	In built-up area	Outside built-up area
1	Motorway	30 m	50 m
2	Express Road	20m	40 m
3	General Access Road: national	10 m	25 m
4	General Access Road: voivodeship,	8 m	20 m
5	General Access Road: municipality	6 m	15 m

The Landscape Act [2] allows municipalities to adopt local advertising placement rules and gives them the power to set rules, collect charges and create media free zones. The advantage of this procedure is the ability to impose large penalties and the enforcement process is quick. So far, implementing the Administrative Procedure Code has meant that the process of removing illegal advertising or imposing penalties has been so slow that advertisers often pay penalties and continue to make significant profits from illegal advertising, or by using procedural deadlines, drag the administrative process indefinitely.

The drawbacks associated with the Landscape Act are:

- Each municipality (gmina) will adopt different regulations, so that for the driver travelling across the country, roads and their adjacent areas will be governed by different standards;
- Many municipalities may pass legislation without taking into account the need to ensure road safety;
- Road authorities are not party to the proceedings in the process of passing local law, so a resolution of the municipality will be implemented without the opinion of the road authorities;
- Municipalities may allow as many advertisements as possible to maximise their revenue from fees.

Currently, drafts of landscape resolutions are being prepared in Poland. Observations conducted in large cities: Warsaw, Cracow, Lodz and Gdansk indicate that only Gdansk and Lodz have taken traffic safety into account, although to a very limited extent.

Bearing in mind the above, it should be noted that advertising regulations in Poland are substantial but insufficient, scattered and difficult to enforce. Some rules do not exist and it is justified to create a comprehensive advertising placement guide.

3 Foreign perspective - the example of Abu Dhabi

The authors have analysed a number of instruction manuals and guides that govern the location of advertising media and how they are displayed. It was decided to present the scope of one of the manuals issued by the Department of Transportation for Abu Dhabi City, [1] which, although not free from minor shortcomings, presents a consistent set of rules for media divisions, locations, parameters and content.

This instruction applies to all advertising visible from the road at a distance of 300 m from its edge. This is very important. In Poland many provisions are valid but the scope of action is very limited. Abu Dhabi adopted the following advertisement classification:

- billboards,

- flags and banners,
- advertising on the carriageway,
- urban furniture,
- mobile advertising,
- other advertising,
- sign boards.

Depending on the type of advertisement, different rules have been introduced, but regardless of the type they cannot:
- present content imitating signalling or traffic signs,
- be mounted on signs or road sign structures,
- be brighter than traffic lights,
- present content that is not completely legible or visible from the road,
- contain more information than can be "read/understood" during two short glances,
- be mounted over carriageways,
- be installed in a central reservation separating carriageways on motorways,
- be mounted near intersections or junctions,
- be assembled in groups,
- contain distracting content,
- have visible light sources,
- have moving and rotating elements,
- emit the sound or smell that the motorists could detect.

Depending on traffic volume and speed limits, so-called "free zones" should be used, where the structure of the advertisement should not be located in a position where there is a risk of vehicle impact that may run off the road. The minimum distance is 3 m and the maximum is 9 m.

The advertisement exclusion zones have been used in this manual, i.e. it is recognised that in some areas the advertisements should be eliminated to ensure that motorists can focus on driving tasks. Exclusion zones were used in the area of intersections (Fig. 1), roundabouts, junctions and exits. The exclusion buffer is calculated differently at the approach to the collision point (zone B) and differently when exiting the collision point (zone A). When exiting the collision zone, ads cannot be located within a distance dependent on speed and the number of units of information contained in the ad. The more information and the higher the speed, the longer the ad-free stretch should be. An ad-free buffer at the approach to the collision point depends only on speed. For example at 50 km/h the buffer is 42 m, for 100 km/h it is 83 m.

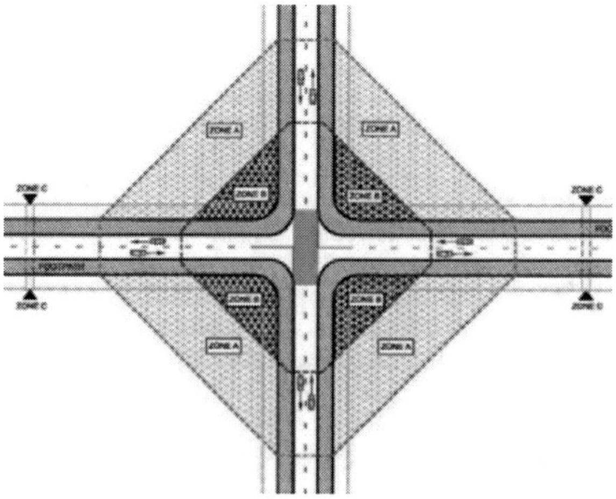

Fig. 1. Cross Intersection Advertisement – free Buffer Zone [1].

The permit is for a specific time and does not exceed 3 years. The application process as presented in Fig. 2 is also noteworthy. The predetermined application is checked for national and local regulations. In case the application is invalid the applicant has 30 days to amend the application. If the application is submitted correctly and the parameters of the advertisement comply with all the regulations, a clerk performs a check in the field. Once the application complies with the site requirements, the terms and conditions and length of contract with the advertiser are determined. The permit is for a specific time and does not exceed 3 years.

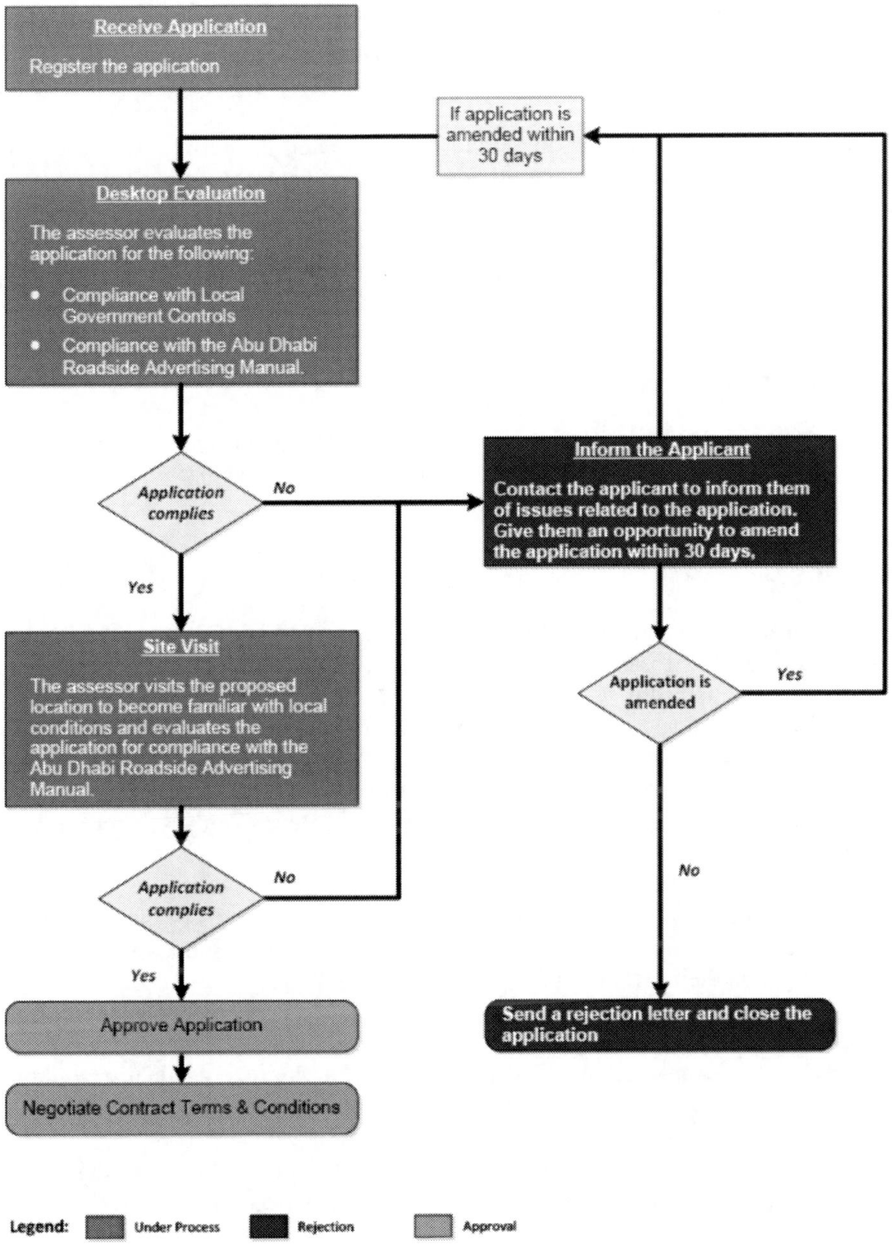

Fig. 2. Procedural Flowchart for Permit Application Review [10].

4 Conclusions

The authors are aiming to develop their own instruction guide for locating and displaying advertisements on national roads. Ultimately, it is recommended to regulate the location and display of advertising on all roads in Poland in order to achieve a high level of traffic safety. Work is underway to assess hazards resulting from driver's distraction caused by roadside advertising [10][11]. In addition, a review of Polish regulations and foreign instructions is conducted. According to the authors, the following principles should constitute the basic principles for the development of instructions regulating the placement and the form of advertising on roads in Poland:

- Advertising does not jeopardize the level of road safety by rule. Advertising can pose a risk of reducing traffic safety if it is located where drivers should focus primarily on driving tasks or where the form of the advertisement (size, lighting or content) will draw the driver's attention for too long.
- The spatial extent of the regulation should not be related to the roadway boundary. The scope of the contract should include all advertisements that are visible from the road and which may affect motorists.
- If motorists are exposed to advertising, its content should be governed with the same approach as road signs. When advertising is displayed, its form and content should be such that the motorist will read and comprehend it within 2-3 seconds.

Acknowledgements

Research conducted under a project implemented within RID programme, funded by the Polish National Centre for Research and Development and the General Directorate for National Roads and Motorways in Poland, for the years 2016-2019.

References

1. Roadside advertising manual, Document No: AD-R-03 First Edition (Abu Dhabi, 2013)
2. Act of 24 April 2015 amending certain acts on strengthening the tools for landscape protection, OJ 2015 Pos. 774 (Poland, 2015)
3. Guideline: Roadside Advertising Guide, Edition 1.2 (Departament of Transport and Main Roads, Australia, 2013)
4. The South African National Roads Agency limited repeal of certain regulations and making of regulations (The South African National Roads Agency, 2000)
5. European Agreement on main international traffic arteries (AGR) (Poland, 1985)
6. Act of 21 March 1985 on Public Roads, OJ 2007 No. 19 Pos. 115 (Poland, 1985)
7. Regulation of the Minister of Infrastructure and Development of 20 October 2015 on the technical conditions to be met by the crossings of railway lines and railway siding with roads and their location, pos. 1744 (Poland, 2013)
8. Act of 20 June 1997 Road Traffic Act no. 98 pos. 3113 (Poland, 1997)
9. Convention on Road Signs and Markings (Vienna, 1968)
10. T. Mackun, K. Jamroz, J. Żukowska, J. Rychlewska, Zagrożenia dla uczestników ruchu drogowego stwarzane przez reklamy widziane z drogi, Drogownictwo. **4–5** pp. 134–141 (2015)
11. K. Jamroz, T. Mackun, J. Rychlewska, J. Wachnicka, Reklamy widziane z drogi: pomagają czy przeszkadzają kierowcom?, in: Cracow Road Saf. Days, (2015)

MATEC Web of Conferences 122, 03006 (2017) DOI: 10.1051/matecconf/201712203006
GAMBIT 2016

Social attitudes towards roadside advertising

Anna Olejniczak-Serowiec [1,*], *Norbert* Maliszewski[1], and *Kinga* Ziętek [1]

[1]University of Warsaw, Faculty of Psychology, Warsaw, Poland

Abstract. Public opinion is not always consistent with expert opinion. A nationwide CAWI research project was conducted in order to verify Polish drivers' opinion on roadside advertising and their experience concerning distraction and loss of situational awareness. Research results show that although many drivers have experienced partial loss of situational awareness, most of them are not convinced of the need to restrict roadside advertising. Nonetheless, on the basis of the results, we can identify the most distracting characteristics of roadside advertising as well as socially acceptable rule changes.

1 Introduction

The idea behind advertising is to have the message memorized or at least acquired by the addressee – a potential client. On the other hand, the effectiveness of driving may decrease as a result of distraction. This leads to the question as to how roadside advertisements influence road safety. Statistical analysis of the number of accidents and behaviour changes reveal some decrease in driver effectiveness in the presence of roadside advertising [1,2], however, other data suggest the contrary [3]. Driving simulator studies show that in the context of roadside advertising, drivers show more behaviours resulting from inattention [4], memorize traffic signs less effectively and feel more tired [5–7], but in each of the studies, some other indicators of driving effectiveness remain unaffected.

As the pure presence of the roadside advertisement does not make a consistent difference in driving behaviour, the question is which roadside advertisement characteristics makes it distracting.

As far as technical characteristics (e.g. light emission, location) are concerned, there is quite a lot of data available [2,7–9]. The influence of the content of roadside advertising on road safety, however, is less described in the literature. Simulator studies accompanied by eye tracking methods prove that advertisements (i.e. information signs with service points' logotypes) are more distracting than traffic signs[10]. Additionally, their distractive value increases with the complexity of the sign (i.e. the number of logotypes) [11]. Another perspective on the complexity of advertisement content was adapted by Scheiber's team [12], according to whom, the danger from roadside advertising occurs when it consists of eight or more words.

Although, the aforementioned studies are a voice in the discussion on the role of advertising content for road safety, they remain focused on technicalities rather than the real substance of the stimuli.

Research on advertising mechanisms proves that the emotional load of advertising influences the addressee's cognitive processes [13]. Consequently, it seems reasonable to hypothesize that the emotional load of roadside advertising may influence driver behaviour

* Corresponding author: anna.olejniczak@psych.uw.edu.pl

and result in lower road safety. This problem became central to Megias' research. In his studies, the influence of negative emotional load turns out to be particularly significant for driving drivers' attention away from the road, but also for decreasing reaction time to danger and the number of speeding decisions when yellow light comes up (compared to positive stimuli exposition) [14,15].

Because the emotional message is an important factor in advertising and is also a proven attention distractor, it was decided that more research on roadside advertisement content, especially the emotional load, would be helpful with understanding the role of roadside advertising for road safety.

2 Study I

The study was inspired by research reports which prove that drivers do consciously pay attention to roadside advertisements and are well aware of the distraction it causes [2,4]. It was designed to explore Polish drivers' experience with roadside advertising with reference to particular types of contents used in advertising.

2.1 Participants and procedure

The data come from a nationwide CAWI research project. All the participants were the research panel's voluntary respondents and were rewarded with a certain amount of points, which they may subsequently exchange for rewards.

Of the total of 1,095 participants, 428 were female and 667 - male. Participants' age varied between 18 and 80 (M=41.43; SD=14.29) and their active driving experience - from less than one year (n=64; 5.8%) to 51 years with a mean of 16.77 (SD=11.56) in the later – more experienced group (n=1031; 94.2%). 718 (65.6%) of the participants declared daily use of the car and 234 (21.4%) declared that they drive a car at least once a week. The eyesight condition was also controlled, 39.5% of the participants declared some impairment.

All the participants filled in an online survey. First, they were asked a set of questions concerning demographics, driving experience and eyesight condition. Subsequently, a number of questions were presented concerning their experience with roadside advertising, as well as their experience with advertisements of particular types of content. Seven types of contents were considered in the study: positive emotional load, negative emotional load, sexual, humorous, riddles or curious contents, sales offers and contents which resemble traffic signs (each illustrated with a photo).

All the questions were divided into two types: YES/NO questions designed to diagnose what percentage of respondents has ever experienced the described situation and frequency questions in which respondents answered on a six point Likert scale (from never to always), how often they face the situation, the answer I don't know was also possible.

2.2 Results

Frequency analysis proved that Polish drivers do sometimes pay attention to roadside advertising (M=2.72, SD=1.41) as well as consciously read it (M=2.26; SD=1.44). Among 42.8% of respondents who declared they had experienced a distraction due to roadside advertising, the mean value for attention distraction frequency was M=3.02 (SD=1.28), they also reported some difficulties in road situation understanding (M=2.66; SD=1.41). The frequency analysis is quoted in Table 1. No statistically significant sex differences were identified, except conscious advertisement reading (t=-2.189; p=0.029), which men (M=2.34; SD=1.44) declare more frequent than women (M=2.14; SD=1.43). Driving experience, education level and eyesight condition did not affect the results either. ANOVA analysis revealed a significant difference between different age groups for paying attention to roadside advertisement (F(4)=4.71, p=0.001) and conscious reading (F(4)=6.99, p<0.001). Post hoc analysis with NIR test shows that drivers over 54 years scored significantly lower in these questions (M=2.37; SD=1.42; M=1.85; SD=1.41) than all the other age groups with mean values varying from 2.74 to 2.87 for paying attention to advertisements and from 2.25 to 2.47 for conscious reading.

Table 1. Frequency analysis for attention distraction, conscious reading of roadside advertisements and difficulties in situation understanding.

	Don't know	Never	Very rarely	Rather rarely	Neiter rarely nor often	Rather often	Very often	Always
Paying attention	2,8%	5,7%	15,9%	20,5%	24,3%	22,5%	6,1%	2,3%
Conscious reading	2,9%	11,5%	20,7%	22,6%	22,6%	13,5%	5%	1,2%
57,2% of respondends have never experienced attention distraction from roaside advertisement; the data below desribe the remaining 42,8% (n=469)								
Attention distraction	1,9%	0,4%	11,9%	24,3%	24,5%	25,4%	9%	2,6%
Situation understanding difficulties	2,8%	6,8%	14,9%	22,4%	23,5%	20,9%	7,9%	0,9%

T-test analysis for dependent samples revealed statistically significant differences in a subjectively evaluated frequency of attention distraction of different content. Mean values for the question "How often do you happen to be distracted by…" are presented in Table 2, and t-test values are presented in Table 3. Group comparison between men and women revealed that women perceive emotionally negative advertisements as more distractive than men (t=4.427; p<0.001; M=3.15 and M=2.43), a similar difference occurs for riddles (t=1.970; p=0.049; M=3.01 and M=2.81). Males find sexual content mores distractive than women (t=-3.324; p=0.001, M=3.02 and M=2.67). When age and length of driving experience were taken into account, a negative correlation was observed, however emotionally negative content and sales offers did not fit the trend. The values of Pearson correlations are presented in Table 4.

Table 2. Frequency of distraction experienced from advertisements with different content.

	Positive	Negative	Sexual	Humorous	Resembling road signs	Sales offer	Riddles
Mean	2,46	2,88	2,88	2,77	2,78	2,37	2,89
Standard deviation	1,44	1,57	1,66	1,49	1,54	1,55	1,54

Table 3. Differences between declared frequency of distraction experienced from advertisements with different content.

	Positive	Negative	Sexual	Humorous	Resembling road signs	Sales offers	Riddles
Positive		t= -11,596***	t=-9,752***	t=-6,380***	t=-8,310***	t=2,183*	t=-10,71***
Negative	t= -11,596***		t= -0,345	t= 5,653***	t= 2,456*	t= - 11,757***	t= 0,024
Sexual	t=-9,752***	t= -0,345		t=5,484***	t=2,271*	t=11,692***	t=0,024
Humorous	t=-6,380***	t= 5,653***	t=5,484***		t=-3,382***	t=7,528***	t=-6,554***
Resembling road signs	t=-8,310***	t= 2,456*	t=2,271*	t=-3,382***		t=10,414***	t=-2,632**
Sales offers	t=2,183*	t= - 11,757***	t=11,692***	t=7,528***	t=10,414***		t= -13,270***
Riddles	t=-10,71***	t= 0,024	t=0,024	t=-6,554***	t=-2,632**	t= -13,270***	

* $p < 0{,}05$ ** $p < 0{,}01$ *** $p < 0{,}001$

Table 4. Age and driving experience correlation with declared frequency of distraction experienced from advertisements.

	Positive	Negative	Sexual	Humorous	Resembling road signs	Sales offers	Riddles
Age	r= -0,108***	r=-0,045	r=-0,093**	r= -0,121***	r=-0,080**	r=-0,004	r=-0,080**
Length of driing experience	r=-0,101**	r=-0,103**	r=-0,081**	r= -0,112***	r= -0,120***	r= -0,081**	r= -0,119***

* $p < 0{,}05$ ** $p < 0{,}01$ *** $p < 0{,}001$

3 Study II

The aim of the study was to verify the impact of emotionally loaded advertisements on attention focus [16].

3.1 Participants and procedure

The study was conducted on 50 (23 female and 27 male) Polish amateur drivers aged from 21 to 51 (M=27, SD=5.72). All of them held their driving license for at least one year (M=7.8; SD=5.72), the mean amount of yearly driven kilometres was 9907 (SD=15161).

A two-task attention test was constructed. The stimuli took the form of a Power Point presentation consisting of a simplified version of Wittenborn test cited after Zomeren and Brouwer [17]. The task consisted of three digits located centrally on the slide. In three corners of the slide, 5-9 simple geometric figures were located, the number of figures was chosen according to the capacity of working memory [18]. The fourth corner was used to display a picture, which was either an advertisement (4 items emotionally positive, 6 – negative, 10 – neutral) or a non-advertisement emotionally neutral picture (10 items). Each participant was presented 3 training and 30 test slides, single exposition lasted 2 seconds. After each slide, they were asked whether the digits were arranged in ascending or descending order and how many triangles they saw in the slide. The amount of correct answers (scores ranged 0-1) served as a performance measure.

3.2 Results

ANOVA analysis did not reveal any statistically significant differences between different stimuli conditions for the simplified Wittenborn task ($F(3)=0.99$; $p=0.39$). For the figures counting task, some effects of conditions were identified ($F(3)=3.55$; $p=0.01$). Post hoc analysis with T-test for dependent samples revealed that the test scores in positive advertisement condition were significantly lower than in negative and non-advertisement neutral picture condition. The difference between positive and neutral advertisements did not reach $p<0,05$ significance level, however, a statistical trend in the same direction was present ($p=0,066$). Mean values for the figure counting test are presented in Table 5, and t-test values, are presented in Table 6.

Table 5. Figure counting test scores in the context of different advertisement content.

	Positive	Negative	Neutral	Neutral non-advertisements
Mean	0,58	0,65	0,65	0,68
Standard deviation	0,27	0,26	0,21	0,18

Table 6. Advertisement content and attention test results.

	Positive	Negative	Neutral	Neutral non-advertisements
Positive		t=2,158*	t=1,881	t=2,953*
Negative	t=2,158*		t=0,04	t=0,913
Neutral	t=1,881	t=0,04		t=1,219
Neutral non-advertisements	t=2,953*	t=0,913	t=1,219	

* $p < 0,05$

4 General Discussion

Consistently with previous research, roadside advertisements attract attention of most of the drivers, with more than a quarter of them declaring that it happens often, and only 5.7% that they never pay attention. Even more compelling, one in five drivers declares frequent conscious reading of the advertisements' content while driving. Some contents of roadside advertisements are subjectively more distracting than others. It seems, that the most subjectively distracting ones are those with emotionally negative contents, sexual contents, and riddles. Experimental data, however, revealed something opposite to the subjective feeling, namely the negative influence of emotionally positive advertising content on attention test performance, though it seems to affect attention to peripherally presented stimuli only (essential for maintaining situational awareness [19]). One thing remains clear which is the fact that emotional value of advertisement content affects drivers' attention which leads us to a conclusion that further experimental research on drivers' attention detachment from the road in the presence of advertisements of different emotional value might be beneficial.

Age-involving analyses justify the hypothesis that distraction from roadside advertising may be particularly dangerous for novice drivers who have not yet fully automatized driving operations and thus need to devote more attention to driving. As long as this seems to be the truth on the level of subjective feelings, experimental verification of this hypothesis might be useful.

Acknowledgements

The research and the preparation of this article were funded by the National Centre for Research and Development and the General Director for National Roads and Motorways (DZP/RID/I-33/4/NCBR/2016).

References

1. G. Yannis, E. Papadimitriou, P. Papantoniou, C. Voulgari, A statistical analysis of the impact of advertising signs on road safety, Int. J. Inj. Contr. Saf. Promot. **20** pp. 111–120 (2013). doi:10.1080/17457300.2012.686042
2. A. Smiley, B. Persaud, G. Bahar, C. Mollett, C. Lyon, T. Smahel, W. Kelman, Traffic Safety Evaluation of Video Advertising Signs, Transp. Res. Rec. J. Transp. Res. Board. **1937** pp. 105–112 (2005). doi:10.3141/1937-15
3. P. Izadpanah, R. Omrani, S. Koo, A. Hadayeghi, Effect of Static Electronic Advertising Signs on Road Safety, J. Orthop. Trauma. **28** pp. S33–S36 (2014). doi:10.1097/BOT.0000000000000101
4. S. Bendak, K. Al-Saleh, The role of roadside advertising signs in distracting drivers, Int. J. Ind. Ergon. **40** pp. 233–236 (2010). doi:10.1016/j.ergon.2009.12.001
5. M.S. Young, J.M. Mahfoud, N.A. Stanton, P.M. Salmon, D.P. Jenkins, G.H. Walker, Conflicts of interest: The implications of roadside advertising for driver attention, Transp. Res. Part F Traffic Psychol. Behav. **12** pp. 381–388 (2009). doi:10.1016/j.trf.2009.05.004
6. T. Horberry, J. Anderson, M.A. Regan, T.J. Triggs, J. Brown, Driver distraction: The effects of concurrent in-vehicle tasks, road environment complexity and age on driving performance, Accid. Anal. Prev. **38** pp. 185–191 (2006). doi:10.1016/j.aap.2005.09.007
7. J. Edquist, T. Horberry, S. Hosking, I. Johnston, Effects of advertising billboards during simulated driving, Appl. Ergon. **42** pp. 619–626 (2011). doi:10.1016/j.apergo.2010.08.013
8. D. Belyusar, B. Reimer, B. Mehler, J.F. Coughlin, A field study on the effects of digital billboards on glance behavior during highway driving, Accid. Anal. Prev. **88** pp. 88–96 (2016). doi:10.1016/j.aap.2015.12.014
9. D. Crundall, E. Van Loon, G. Underwood, Attraction and distraction of attention with roadside advertisements, Accid. Anal. Prev. **38** pp. 671–677 (2006). doi:10.1016/j.aap.2005.12.012
10. D. Kaber, C. Pankok, B. Corbett, W. Ma, J. Hummer, W. Rasdorf, Driver behavior in use of guide and logo signs under distraction and complex roadway conditions, Appl. Ergon. **47** pp. 99–106 (2015). doi:10.1016/j.apergo.2014.09.005
11. Y. Zhang, E. Harris, M. Rogers, D. Kaber, J. Hummer, W. Rasdorf, J. Hu, Driver distraction and performance effects of highway logo sign design, Appl. Ergon. **44** pp. 472–479 (2013). doi:10.1016/j.apergo.2012.10.009
12. F. Schieber, K. Limrick, R. McCall, A. Beck, Evaluation of the Visual Demands of Digital Billboards Using a Hybrid Driving Simulator, Proc. Hum. Factors Ergon. Soc. Annu. Meet. **58** pp. 2214–2218 (2014). doi:10.1177/1541931214581465
13. K. Oatley, J.M. Jenkins, Zrozumieć emocje (Wydawnictwo Naukowe PWN, Warszawa,

2003)

14. A. Megías, L.L. Di Stasi, A. Maldonado, A. Catena, A. Cándido, Emotion-laden stimuli influence our reactions to traffic lights, Transp. Res. Part F Traffic Psychol. Behav. **22** pp. 96–103 (2014). doi:10.1016/j.trf.2013.09.017
15. A. Megías, A. Maldonado, A. Cándido, A. Catena, Emotional modulation of urgent and evaluative behaviors in risky driving scenarios, Accid. Anal. Prev. **43** pp. 813–817 (2011). doi:10.1016/j.aap.2010.10.029
16. K. Ziętek, Unpublished master thesis: Wpływ ładunku emocjonalnego reklam na koncentrację uwagi wzrokowej, University of Warsaw, 2016
17. A.H. Zomeren, W.H. Brouwer, Clinical neuropsychology of attention (Oxford University Press, 1994)
18. A. Baddeley, D. Chincotta, L. Stafford, D. Turk, Is the word length effect in STM entirely attributable to output delay? Evidence from serial recognition, Q. J. Exp. Psychol. Sect. A. **55** pp. 353–369 (2002). doi:10.1080/02724980143000523
19. A. Tarnowski, Eye Movements as the Window into Visual Attention Mechanisms and Situation Awareness, Polish J. Aviat. Med. Psychol. **18** pp. 45–50 (2012)

MATEC Web of Conferences **122**, 03007 (2017) DOI: 10.1051/matecconf/201712203007
GAMBIT 2016

Automatic road traffic safety management system in urban areas

Jacek Oskarbski[1*]

[1]Gdansk University of Technology, Faculty of Civil and Environmental Engineering, Gdansk, Poland

Abstract. Traffic incidents and accidents contribute to decreasing levels of transport system reliability and safety. Traffic management and emergency systems on the road, using, among others, automatic detection, video surveillance, communication technologies and institutional solutions improve the organization of the work of various departments involved in traffic and safety management. Automation of incident management helps to reduce the time of a rescue operation as well as of the normalization of the flow of traffic after completion of a rescue operation, which also affects the reduction of the risk of secondary accidents and contributes to reducing their severity. The paper presents the possibility of including city traffic departments in the process of incident management. The results of research on the automatic incident detection in cities are also presented.

1 Introduction

Transport systems are a key element in meeting basic social needs related to population mobility and supplies. They facilitate the development of economic activity and contribute to upgrading the quality of life. Society expects a high level of reliability of transport while traveling to work, schools and for recreation. Today, however, it is already known that certain changes in transport management are necessary to tackle such urgent problems as traffic congestion, environmental pollution, traffic safety or social risks, at the same time ensuring sufficient mobility in the future. The development of Intelligent Transportation Systems (ITS) as supplementary to long-term measures in transport policy offer hope for at least a partial solution to the problems encountered, including in particular the problem of transport safety. Consequently, many cities in developed countries introduce intelligent systems of transportation management. The intensified implementation of ITS services has also been observed in Polish towns and on rural roads for several years now. Traffic services are not the sole beneficiaries of such ITS services. New software and hardware solutions are also applied by institutions participating in rescue operations (medical services, fire brigades) and preventive actions (e.g. those related to speed control). Systems used by road authorities as well as emergency and preventive services provide detailed data on traffic incidents and other big data which can be used at planning and operational levels of security management. These data are also a potential resource that could be the basis for research aimed at improving and developing new methods and means to upgrade traffic safety. Changes taking place through technological development, that manifest themselves in an ever wider use of automation in transportation processes, justify the revision of the current approach to traffic safety

* Corresponding author: joskar@pg.gda.pl

management in order to strengthen the integration of services and to develop procedures for their cooperation.

2 Traffic safety management system

2.1 System components

According to the definition the "transportation system" is a system of technical, organizational and human resources, combined with one another in such a way that it can efficiently carry out the traffic of people and goods in space and time [1]. In the system of traffic safety management, which is inextricably linked to the transportation system, we can identify many interrelated and mutually interacting elements:

- structures of organization of traffic safety management services, as well as legal bases and procedures that determine the organization, competences and collaboration of such services,
- tools influencing the improvement of traffic safety management at the strategic level (e.g. regional, local and sectoral programs of traffic safety management, including monitoring of their implementation) and the operational level (current monitoring and supervision),
- broader support tools for system management (e.g. databases, information systems, expert systems, guidelines and examples of good practice, ITS services) based on scientific research,
- management methods for the whole system and its selected elements (including risk management).

The National Road Safety Program for 2013-2020 [2] indicated that the process of improving traffic safety requires the use of up to three consecutive and interlinked components: institutional management functions, specific actions (interventions) and outcomes. Basic functions of institutional management included coordination, legislation, financing and provision of resources, promotion and communication, monitoring and evaluation, as well as research, development and knowledge transfer.

The diagnosis of the existing system of safety management carried out in the framework of the National Road Safety Program has shown that each of these functions needs to be improved in terms of implementation of measures included in the Program. These functions are performed in different proportions depending on the particular institution and the level of public administration. They highlighted the need to improve organizational structures of traffic safety and coordination with the National Road Safety Council as a leading institution, as well as with Regional Road Safety Councils as the real leaders of the region – with the support of research institutions. Regional Road Safety Councils should perform management functions, among others, in terms of horizontal coordination - between regional structures of the Police, Road Transport Inspectorate, State Fire Service, the General Directorate for National Roads and Motorways, regional roads authorities, county roads authorities, the school superintendent, local NGOs and local communities, as well as through vertical coordination - with the National Road Safety Council and county road safety councils [2]. The representation of such a broad body in the works of road safety councils provides an opportunity to develop conclusions regarding necessary legislative changes and investigating the needs of cooperation in the management of traffic safety at the level of use of technology and databases as well as their integration, which may result in modification and development of new procedures for cooperation between the services. An example of new work on the introduction of a uniform system of monitoring and communication is the Polish Road Safety Observatory, functioning at the Motor Transport Institute, and emerging regional observatories. Equally important is also gathering data on traffic parameters that – when compared with data on traffic incidents - can be helpful in developing methods for improving traffic safety. Such data can be collected, for instance, in ITS Regional Laboratories. The structures of these databases and their linkages should be developed. It is also important to ensure universal access to data. The activities of local governments working on solutions in the field of "Smart City" can be helpful in this respect. The result of these activities should be the introduction of traffic safety and knowledge transfer.

2.2 Operational traffic safety management

The system of transport suffers disruptions every day. Transport systems become less reliable (which in particular affects the reliability of travel time within a transport network) [3] when undesired incidents take place. These include dangerous incidents (accidents and collisions) and incidents (such as a vehicle becoming stationary due to a breakdown or driver indisposition, objects on the road, failure of parts of road infrastructure) [4]. This suggests that a systemic solution is required to help restore normal traffic conditions faster after the accident scene is cleared and minimize travelers' time lost. The objective is also to reduce the risk of secondary events (by keeping motorists informed about the incident and controlling speed via Variable Message Signs) and accident severity (by reducing the time for emergency services to get to the scene and attend to the casualties). With the deployment of ITS services in Poland the work of the services can now be intensified and integrated as part of operational road safety management. The architecture of the Tri-City's Integrated Traffic Management System TRISTAR provides for the delivery of an operational road safety management system as depicted in Fig. 1.

Traffic Incident Management (TIM) is a systematic, planned and coordinated use of resources to reduce the impact of incidents, and improve the safety of motorists, crash victims and incident responders and to restore normal traffic conditions after the occurrence of an incident is detected on the road. Using its detection capability, the Monitoring and Traffic Surveillance System supports the collection of data about traffic parameters, weather conditions, travel times on specific street sections. Once processed they can be used by Traffic Safety Management System modules. The Automatic Supervision of Drivers Behavior Module is designed to detect road traffic offences such as exceeding momentary speed and average speed on a road section and driving over the STOP line on a junction or pedestrian crossing on a red light for the particular direction. At present, the modules collect offence statistics, a useful source of information for road authorities when they take steps to improve traffic layout and for the police when they take preventive action. The data can be used to send complete sets of offence information to a central traffic enforcement body. Before drivers can be punished for offences they have committed, the law and organizational arrangements must be revised and verified. The next task of the module is to pass on travel time information for use in the Traveler Information System (drivers can access the information via VMS and website) [5].

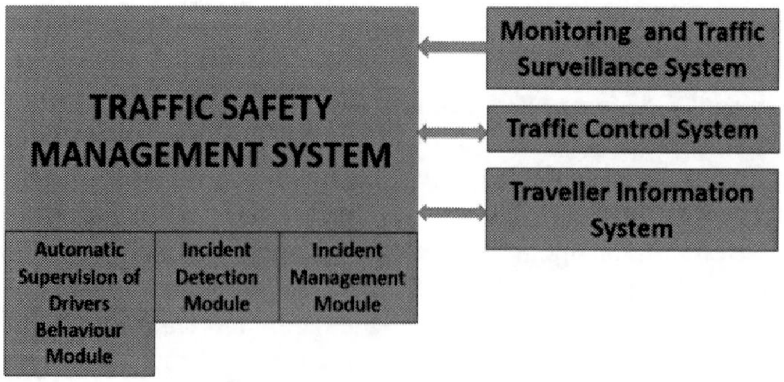

Fig. 1. Traffic safety management in TRISTAR system.

In the case of the Incident Detection Module - a pilot implementation of the first element of the module has been completed. The project enables automatic detection of traffic incidents on a few selected street sections in the Tri-City (a dual carriageway with 2 lanes in each direction, Polish class of street - GP), e.g. on Gdynia's Kwiatkowski Route. The main element of the data processing module are algorithms that analyze available data to detect incidents. The module uses data from Bluetooth and Wi-Fi scanners that detect Bluetooth or Wi-Fi devices (mobile phones, smartphones, tablets, computers and in-vehicle devices to support hands-free systems – e.g. in-car speaker phones). The data collected by each pair of scanners (located at the beginning and end of the road section) are used to calculate travel times on defined distances between scanners. An algorithm based on Kalman filter is used to verify the data transmitted from the network of scanners [6]. The incident detection algorithm checks the changes in travel time between measuring points every minute and when a sudden and unjustified change in travel time occurs, a notification is sent to the system operator with proposed information to be displayed on VMS [7],[8]. This is how the Incident Management Module goes into operation. It can start a traffic management strategy if there is an incident. It uses the Traveler Information System (warnings displayed on VMS about an incident, detours, road closure, speed management) and the Traffic Control System (change in signalization programs in response to a change in traffic distribution on the street network).

3 Effects of using ITS services for operational road safety management

When there is a road accident, the victim may suffer serious injury leading to their death or disability, unless help is provided without delay. A fast response by emergency services may substantially increase the chance of survival and reduce long-term consequences of the trauma. According to the principle of the "golden hour" the lives of 20-40% of seriously injured casualties can be saved, if they receive hospital treatment within 60 minutes from the event. They are the more likely to live, the sooner they receive first aid at the scene (within "ten golden minutes") before they are taken to hospital. European research estimates that transport telematics can reduce the response time and intervention of emergency services by as much as 30% [9]. When an incident is detected and verified, a strategy is activated which should ensure that a rescue team is quickly put together and the necessary equipment is made available. The response to an incident includes dispatching services to the scene. The steps should also include informing motorists once it is certain that the accident happened. Fast response to incidents means that the appropriate services must take action immediately. To speed up the process of detecting dangerous incidents, automatic incident methods can be used. If detected early, the incident will have a shorter duration and getting help to the casualties will take less time. As a result, accident severity will be reduced and the time of exposure to risk of secondary incidents will be minimized. Travelers' time lost will also be minimized [10],[11]. Other benefits include a reduction in serial and secondary incidents (up to 29%), drivers' working time, fuel consumption and emissions [12],[13]. The majority of the city algorithms were developed to detect incidents on sections of the arterial road between junctions and some of them support detection on junctions. Under the CIVITAS DYN@MO project algorithms were developed using time series and artificial neural networks that can detect an incident on signalized junctions [8], [14]. Fig. 2 shows an example of an incident detection using the algorithms – the FiltrLP model. The model signalizes differences in current trend, speed measured at a control point PK_0 (traffic measurement station) and the "normal" trend, i.e. the trend on a day that is similar to the day in question on which there were no incidents.

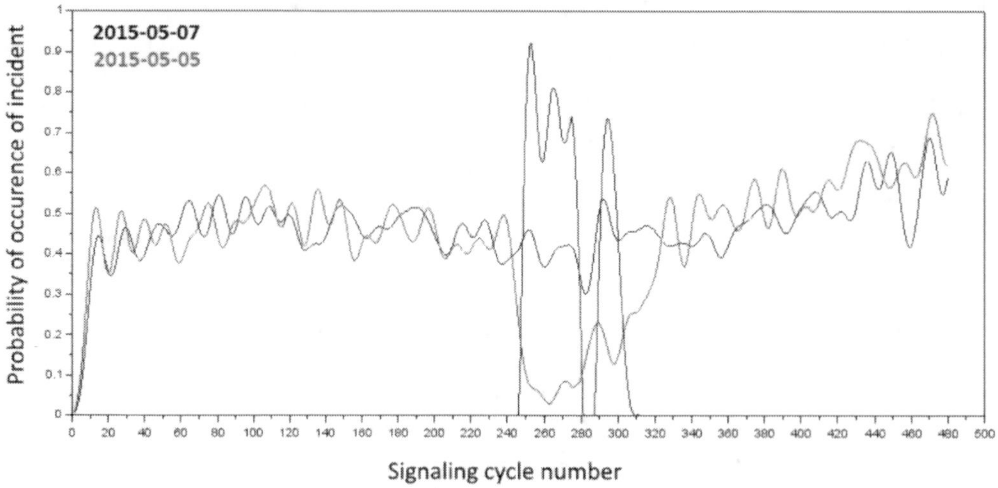

Fig. 2. Answer of the model (blue) - the probability of occurrence of the incident on 2015-05-05 established on the basis of the difference in trends ranks PK_0_V on 2015-05-07 (black) and 2015-05-05 (red).

4 Conclusions

With new ITS services deployed in Polish cities and on national roads in recent years, it is possible to improve the traffic safety system at the strategic (planning) and operational level. There is an increasing number of databases that store big data from integrated transport management systems. These should be integrated and made available as Open Data to improve ITS services and applications to stimulate the emergence of new ones. If the data from the databases can be analyzed, new methods and tools for analyzing road safety can be created. The system of safety monitoring in real time and over extended periods can improve as a result. With data stored in a single system, they can be integrated and used in applications for exchanging data between stakeholders. Once processed, the data can be presented on an information platform (it would be useful to integrate IT systems and databases of the relevant services to ensure data security and controlled access to sensitive data). An important step towards improving the flow of information and notifying the services is the National Access Point deployed by the General Directorate for National Roads and Motorways. ITS resources can also be used to support cooperation between emergency and preventive services with traffic control. This, however, will require new procedural and equipment standards (to ensure equipment compatibility and communications) at the national level including the local specificity and stakeholders involved in incident management and inclusion in the system of Traffic Management Centers. The first procedures of cooperation between emergency services and road authorities were developed as part of Rescue Plans for motorways and express roads. There is opportunity to disseminate these procedures as the National Traffic Management System (KSZR) is developing. The procedures should also be implemented in cities that already have integrated ITS solutions. One of the main steps towards a successful improvement of the road safety management system is developing clear legal bases and verifying the existing ones.

Acknowledgements

Scientific study financed from funds for science in the years 2013-2016 granted to an international project co-financed within the CIVITAS DYN@MO project.

References

1. M. Jacyna, Wybrane zagadnienia modelowania systemów i procesów transportowych (Oficyna Wydawnicza Politechniki Warszawskiej, 2009)
2. National Road Safety Programme for 2013-2020 (Polish National Road Safety Council, Warsaw, Poland, 2013)
3. J. Oskarbski, K. Jamroz, Reliability and Safety as an Objective of Intelligent Transport Systems in Urban Areas, J. KONBiN. **2** pp. 59–70 (2015)
4. K. Jamroz, A. Kadziński, A. Szymanek, K. Chruzik, L. Gucma, J. Skorupski, Integracja metod zarządzania ryzykiem w transporcie, in: Zintegr. Syst. Bezpieczeństwa Transp. Tom 2 Uwarunk. Rozw. Integr. Syst. Bezpieczeństwa Transp., (2009)
5. J. Oskarbski, M. Zawisza, M. Miszewski, Information System for Drivers Within the Integrated Traffic Management System - TRISTAR, in: Commun. Comput. Inf. Sci. Vol. 531, pp. 131–140 (2015)
6. J. Barcelo, L. Montero, M. Bullejos, O. Serch, C. Carmona, A Kalman filter approach for exploiting Bluetooth traffic data when estimating time-dependent OD matrices, J. Intell. Transp. Syst. Technol. Planning, Oper. **17** pp. 123–141 (2013)
7. J. Oskarbski, Perspectives of Telematics Implementation in Tri-City Transport Systems Management and Planning, in: Commun. Comput. Inf. Sci. Vol. 239, pp. 233–240 (2011)
8. J. Oskarbski, M. Zawisza, K. Żarski, Automatic incident detection at intersections with use of telematics, in: Transp. Res. Procedia Vol .14, pp. 3466–3475 (2016)
9. M. McDonald, H. Keller, J. Klijnhout, V. Mauro, R. Hall, A. Spence, C. Hecht, O. Fakler, Intelligent Transport Systems in Europe. Opportunities for Future Research, World Sci. (2006)
10. K.J. Button, D.A. Hensher, L. Schintler, Handbook of Transport Systems and Traffic Control (Pergamon, 2001)
11. J. Oskarbski, Automatyzacja zarządzania zdarzeniami drogowymi, Drogownictwo. **4** pp. 153–156 (2012)
12. G.L. Chang, S. Rochon, Performance Evaluation and Benefit Analysis for CHART in Year 2007, Final report (2007)
13. Analytical procedures for determining the Impacts of Reliability Mitigation Strategies, SHRP2 Reliability Research, Report S2-LO3-RR-1 (Transportation Research Board, 2013)
14. J. Oskarbski, K. Palikowska, K. Żarski, Systemy automatycznego wykrywania zdarzeń niepożądanych w miastach, Pr. Nauk. Politech. Warsz. Transp. **114** pp. 245–254 (2016)

Systematics of Intelligent Transport Systems Services

Jacek Oskarbski[1,*], *Tomasz* Marcinkowski[1], *Krystian* Mówiński[1], and *Karol* Żarski[1]

[1]Gdansk University of Technology, Faculty of Civil and Environmental Engineering, Gdansk, Poland

Abstract. Recent years have seen a more intensified deployment of Intelligent Transport Systems (ITS) in Polish cities. Work is also underway on the implementation of ITS on the national roads within the National Traffic Management System (KSZR). The research project RID-4D (part of Road Innovations), is designed to fill the gap, which is the lack of systematics of ITS services in Poland. The paper presents a proposal of ITS services systematics, which was developed based on an analysis of key international standards. FRAME architecture and American standards were the basis for the preparation of a standard scheme of ITS services. The results of the work on the KSZR architecture as well as local and regional architectures of urban systems were also taken into account.

1 Introduction

The main objective of the RID 4D research project is to investigate how ITS services can improve road safety. Because Poland does not have an ITS architecture, ITS services first had to be identified, systematised and grouped to support further analyses of how single services or groups of services can influence road safety and efficiency in Poland. Existing deployments in Polish cities (such as ITS functions and services which are part of the Tri-City TRISTAR system) [1] and the documentation of the planned KSZR had to be taken into account. Before proceeding with the ITS services systematics, key publications on ITS architecture and services were analysed [2], [3], [4]. The most important baseline document for preparing the systematics was FRAME, the European architecture paper. The names of the services were considered both from the perspective of user needs (USER NEEDS) and the functionality of the service (FUNCTIONS). Importantly, the planned functional structure of the KSZR was analysed [5]. The end result is a model systematics of ITS services developed in collaboration with the General Directorate for National Roads and Motorways (GDDKiA). Each service comes with a detailed description to ensure clarity and functionality of the particular ITS services.

This article analyses the literature, domestic and international experience on ITS classification in relation to the services (a synthesis of international experience in the systematics of services, diagnosis of service development and worldwide strategies for service management, synthesis of Polish experience – work of the ministry, GDDKiA, local authorities) and how this may help improve traffic conditions and road safety. Finally, a model systematics of services has been developed (divided into structures, categories and

* Corresponding author: joskar@pg.gda.pl

names) with a description of the functional, logical and physical structure (hardware and teleinformatics).

2 Studies of literature on ITS services systematics

2.1 General definition of an ITS service

The basic problem of the analysis was to define clearly the term "ITS service". According to the Directive of the European Parliament and of the Council 2010/40/EU of 7 July 2010 on the framework for the deployment of intelligent transport systems in the field of road transport and for interfaces with other modes of transport, it was stated that:

"ITS service" means the provision of an ITS application through a well-defined organisational and operational framework with the aim of contributing to user safety, efficiency, comfort and/or to facilitate or support transport and travel operations; where "ITS application" means an operational instrument for the application of ITS [6]. To conduct a more in-depth analysis of the term and gain a better understanding of how ITS services are systematised, several documents were consulted which set out how this is done in the world, Europe and Poland:

- FRAME (Europe)
- KSZR (Poland – based on FRAME)
- ISO standard
- US architecture - US National ITS Architecture

2.2 ITS services in the FRAME architecture

FRAME architecture is now available in its fourth version and covers the following functional areas:

- electronic payment,
- safety, traffic management,
- public transport management,
- driver assistance,
- traveller assistance,
- support for law enforcement,
- fleet management,
- cooperative systems (exchange of information between vehicles and vehicles and infrastructure) [7].

The methodology for designing ITS systems under the European FRAME architecture distinguishes four main stages. For the purposes of the analysis, the first two ones are selected:

1. Identify user requirements [USER NEEDS];
2. Identify the system's functional architecture to ensure that user needs and functions match (taking into account available functions, their delivery and information flows between them) [FUNCTIONAL VIEWPOINTS];
3. Identify the physical architecture by defining sub-systems and modules, their location and flows between them;
4. Identify the architecture of the communications system, i.e. means of information exchange between the different parts of the traffic management system and external entities [2].

As regards the design methodology, the most important element of FRAME is to identify user requirements and the system's functional architecture. For this reason, USER NEEDS and FUNCTIONAL VIEWPOINTS were analysed in great detail in the initial part of this work during the stock-taking of services and then when developing the proposed systematics.

The FRAME architecture contains orderly structural groups/areas for the user need branch and functionality branch which come together at the lowest levels of detail (Fig. 1.). This helps to identify the physical and communications architecture at a later stage.

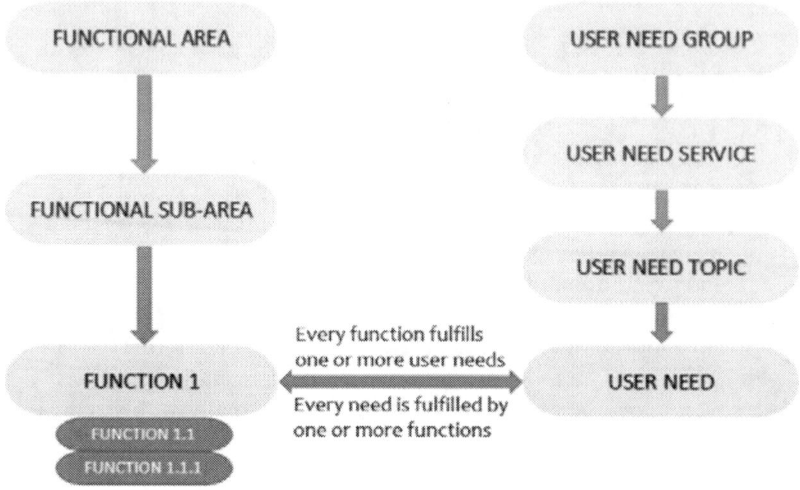

Fig. 1. Relations between user needs and functionalities in FRAME.

According to FRAME a clear-cut definition of an ITS service was not easy due to the specific names of the components of system structures.

FRAME provides the necessary tools for designing ITS architecture which help to define the links between user needs and the functionality of the service. They are called SELECTION TOOL (an application which supports the process of developing ITS architecture for the user) and BROWSING TOOL (an application that allows the user to investigate the structure of the FRAME architecture). The tools and other materials found on the FRAME website helped to identify the links and understand the systemic structure of FRAME which was the basis for creating the model systematics of ITS services.

2.3 ITS service in US architecture

A study of the US's National ITS Architecture was conducted as part of the in-depth analysis. The architecture provides the necessary tools for planning, defining and integrating all intelligent transport systems. It defines:
- The functions (e.g. gather traffic information) that are required for ITS
- The physical entities or sub-systems where these functions reside
- The information flows and data flows that connect these functions into an integrated system.

The architecture is divided similarly to the FRAME architecture. The main component is USER SERVICES and USER SERVICES BUNDLE which categorise services into groups [4].

2.4 ITS service according to ISO

Another helpful source for developing the proposed systematics was service categorisation in the ISO standard 14813-1:2015 [2]. Services here are also arranged hierarchically:
- ITS Service Domain
- ITS Service Group

- ITS Service

The ITS service includes a product or activities that may be supplied to a specific ITS user. This is considered an elementary "building block" of any architecture/implementation.

2.5 ITS service within the National Traffic Management System

The next important step was to see how the new systematics would work with the structure of the National Traffic Management System (KSZR) for Poland's national roads developed by the GDDKiA. The main point of reference for developing the system's service architecture was the architecture of FRAME. The GDDKiA is using a dedicated KSZR website to provide a lot of information about the system's general functional and physical architecture based on the European FRAME architecture. The website also gives a description of implementation modules. Key to the work on the systematics was to refer to the modules [5].

The functional architecture consists of descriptions of the functions and flows between them. This is presented in a table and a graphic representation of the functions and flows. The physical architecture divides the KSZR into implementation modules. There are links between the implementation modules, operators and external stakeholders. The functionalities are also described with reference to FRAME. Treating functions as ITS services turned out to be the right approach for the new services systematics [5].

3 Methodology of work on ITS service systematics

3.1 Stage One: Selecting the level of detail with reference to FRAME

The main point of reference for the new systematics was the FRAME architecture, Europe's leading base material for planning ITS (the KSZR's ITS architecture is an example). Its SELECTION TOOL helps to establish the connections between a need and functionality (activity) to meet that need. Using a database, we can find the details of the relation. The team's main task was to select those components that will best reflect the 3-level detail of service categorisation. While a USER NEEDS analyses showed that it contains the majority of the services, they are designed to provide benefits for the user. This was not a strong enough argument to use in the systematics because despite a clear level of detail, problems could appear later and service functionalities might not be accurately described. A FUNCTIONS oriented approach guaranteed a wide variety of services and a detailed service description. Service categorisation, however, features different levels of detail, which turned out to be a problem. The above approaches were compared and the functional approach was chosen. The problem of the level of detail was solved at a later stage of work on the systematics by grouping analogous functions.

3.2 Stage Two: Expert method while reviewing FRAME functions

The functionality approach meant having to analyse thoroughly all ITS functions (more than 300) of the FRAME architecture and grouping them in a specific level of detail: category (area) of services -> main service -> specific services.

Initially, all services were selected for how they affect road safety and traffic efficiency. Using the expert method a preliminary classification was conducted. This started a discussion on some of the problematic functions involved in data collection; should they be classified as a separate service or as an indispensable component of service delivery. Finally, functions were grouped within a single selected specific service. This is reflected in the tabularised version of the model service systematics.

3.3 Stage Three: Categorisation and division of ITS services

In the final stage a tabular three level division of the service was proposed based on the agreed guidelines. For clarity's sake the names are in two languages to avoid any confusion with the FRAME architecture. In addition, each of the specific services comes with one or several FRAME functions in the original numbering. It was important to do that to ensure that no significant function was left out when presenting the procedure for systematics development.

ID	Category	ENG	ID	Main service	ENG	ID	Specific service	ENG	FRAME function	ID	GDDKiA service
3	**Zarządzanie ruchem**	**Traffic Management**	3.4	Zarządzanie zdarzeniami niepożądanymi	Incidents Management	3.4.1	Wykrywanie zdarzenia	Detect incidents	3.2.12| 3.2.13		-
						3.4.2	Ocena zdarzenia i reagowanie na nie	Classify, identify incidents and devise responses	3.2.11| 3.2.6		-
						3.4.3	Niwelowanie skutków zdarzenia niepożądanego	Provide incidents mitigations to traffic management	3.2.7	4.1	Wdrażanie organizacji ruchu dla zarządzania zdarzeniami
						3.4.4	Informowanie o szczegółach zdarzenia niepożądanego	Send incident details	3.2.14|3.2.8|3.2.9	1.1	Informacja o zdarzeniach drogowych
						3.4.5	Gromadzenie, przetwarzanie i archiwizacja danych o zdarzeniach niepożądanych	Manage store of incident data	3.2.10	3.6	Krajowy Punkt Dostępu

Fig. 2. Fragment of the new services systematics using the example of a main service which is Incidents Management. [8]

In the final stage of work, some names were corrected as necessary and unclassified KSZR services were allocated as appropriate. The figure (Fig.2) shows an example of the new systematics for a main service, i.e. Incidents Management.

To ensure that the services are defined in every detail, each specific service was described on the basis of FRAME original functions, which includes the functional, logical and physical layers. The work has produced a systematics with 10 categories divided into 30 main services and 101 specific services.

4 Summary and further work

Developed to study the effects of ITS services on road safety, the Polish systematics of ITS services was largely based on known ITS architectures. An important part of the work was to decide on the level of detail of the services. The structure of the systematics was to help understand the effects of services and groups of services on road safety and traffic performance. The model systematics takes accounts of the ITS services used in Polish cities and in the KSZR. A review of literature was conducted to understand the effectiveness of the services. The proposed systematics of services was the basis for conducting surveys in local authority and central government institutions that are responsible for infrastructure management, road traffic, public transport, emergency and security services. Companies and NGOs involved in ITS were also surveyed. Thanks to the operational experience of the respondents, suggestions have been made regarding the nomenclature of ITS services included in the systematics. The survey results will be presented in future publications. The developed systematics will help to choose the most relevant services in the next stages of research.

Acknowledgements

The article is part of the project RID-4D (41) "The impact of the use of Intelligent Transport Systems services on the level of road safety" (OT4-4D/ITS-PW-PGIBDiM-WAT), funded by the National Centre for Research and Development (NCBR) and the General Directorate for National Roads and Motorways – GDDKiA (contract no. DZP/RID-I-41/7/NCBR/2016 of 26 February 2016).

References

1. J. Oskarbski, Perspectives of Telematics Implementation in Tri-City Transport Systems Management and Planning, in: Commun. Comput. Inf. Sci. Vol. 239, pp. 233–240 (2011)
2. European Intelligent Transport Systems (ITS) Framework Architecture, www.frame-online.eu, (2016)
3. Intelligent Transport Systems - Reference model architecture(s) for the ITS sector - Part 1: ITS service domains, service groups and services, ISO 14813-1:2015(E) 2-nd ed. (Geneva, Switzerland, 2015)
4. Key concepts of the National ITS architecture - Users Services and User Service Requirements, available at: www.iteris.com/itsarch/documents/keyconcepts/keyconcepts.pdf, (2016)
5. Krajowy System Zarządzania Ruchem, www.kszr.gddkia.gov.pl, (2016)
6. Directive 2010/40/EU of the European Parliament and of the Council of 7 July 2010 on the framework for the deployment of Intelligent Transport Systems in the field of road transport and for interfaces with other modes of transport, (2016)
7. M. Litwin, K. Modelewski, Tworzenie usług dla transportu publicznego w oparciu o europejską Architekturę FRAME, in: VIII Konf. Nauk., (Poznan, Poland, 2011)
8. J. Oskarbski, M. Tomasz, M. Krystian, M. Zawisza, K. Żarski, Systematics of ITS services – periodic report of the project RID-4D: The impact of the usage of Intelligent Transport Systems services on the level of road safety (Gdansk University of Technology, 2016)

MATEC Web of Conferences **122**, 03009 (2017) DOI: 10.1051/matecconf/201712203009
GAMBIT 2016

Preliminary simulation research of driver behaviour in response to outdoor advertisements

Tomasz Targosiński [1,*]

[1]Motor Transport Institute, Warsaw, Poland

Abstract. Advertisements placed near roads can pose a hazard to road safety because they attract the driver's attention. Many attempts to describe and reduce this impact have been done. However, existing regulations treat this problem differently in different countries which demonstrates the significant difficulties in defining and investigating the impact factors of advertisements on driver behaviour. Pilot studies have been done using the eye-tracker and vehicle simulator. Based on the analysis of driver perception in a programmed and repeatable environment an attempt was made to select key factors influencing their attention in terms of the content of advertisements, the size and location in relation to the road and the personality of drivers.

1 Introduction

The basic intention of outdoor advertising is to attract the observer. From the road safety point of view it is antagonistic to the need of road observation. It is obvious that placing advertisements in the field of view of drivers can negatively influence safety. There are a number of factors that describe and affect distraction during driving. But it is not obvious which ones are really important for safety. They include: the size of billboards, their brightness, position regarding driver field of view. But the content and shape of the ad seem to be most important. Driver perception and behaviour is also influenced by human factors such as habit. The question arises: how to define and measure the influence of ads on drivers' attention. National or local governments decide whether a given advertisement should be allowed or prohibited. As a result, there are different regulations regarding advertisements near roads in different countries, starting from total prohibition of advertisements close to the roads to allowing them nearly everywhere. Nevertheless, there is an objective need to properly regulate this area. A lot of times the decisions to place advertising are subjective [1]. In such attempts there are hidden difficulties with "measuring" the content and shape of the advertisement [2–7]. There were also many experimental tests investigating the influence of ads on driver behaviour. Test were conducted with drivers in real traffic and in driving simulator conditions. They reported various results which depend on arbitrarily chosen test parameters such as lane control, time of reaction, scenario, etc. [8–13], Many tests concern different driving parameters in relation to the very simple state of ads or no ads. The difficulty is that researchers usually try to establish whether the advertisement changes any of measurable parameters such as trajectory, speed or driver workload. The choice of content of

* Corresponding author: tomasz.targosinski@its.waw.pl

ads used for the tests is usually subjective with a very general description [4,8,10]. Generally, changes were observed in some measurable parameters but other results showed no changes [12]. The influence of the content of an ad was not tested in any of the known research. It defined in a very general and subjective way, e.g. a negative or positive emotional load [13]. The important question is how big is this emotional load, how to classify it and how different persons react depending on personal characteristics and life experiences. When the number of people is relatively low the results might strongly depend on it.
The tests described below were intended to investigate the influence of the content of an ad on driver attention. It was expected that such influence actually occurs but there was no reference as to how much and how. Therefore the goal was to obtain a general view if such a phenomenon existed and to find directions for further research. The tests were not intended to obtain quantitative values. Because of its preliminary character and not many participants, the statistical analysis would not be very useful and might be misleading because of the small number of test persons. The objective was to detect how much the driver can be attracted by different content, size and placement of ad. The investigators look for general results only, using eye tracking which helps to quickly understand whether the driver was attracted or not by an ad. Future more advanced test with restricted test conditions will be the next step and quantitative findings will be studied for their influence on road safety.

2 The conditions of the experiment

The tests were conducted in a driving simulator AS1200-6. This simulator consists of a passenger cabin positioned on a system of hydraulic actuators. Before the cabin there is a cylindrical screen for projecting digital dynamic images simulating road scenarios. Additionally, there are displays of rear view mirrors. All vehicle controlling devices and cockpit displays are like those in a regular passenger car. This equipment allows for realistic driving in a computer controlled environment. The main advantage of this kind of experiment was the possibility to create situations which are very difficult to obtain in real road conditions because of safety. Also the repeatability of driving scenarios was higher.
During the experiment, areas of driver sight concentration were registered. For this purpose SMI eye-tracking system was used. Heat maps were used which represent absolute gaze duration. The sensor device was in the form of glasses (Fig. 1.)

Fig. 1. SMI eye tracking system in the form of glasses.

Driving with the glasses is similar to real conditions. However, some negative phenomena were observed. Some female drivers and people wearing correction glasses were excluded from the experiment because of the sensitivity of the eye tracking device to long eyelashes and optical glasses. Also the calibration of the device can significantly influence the accuracy of the results. For example, if there is a road or traffic sign close to the advertisement, it can be confusing as to what is the object of the driver's interest. If the driver touches the glasses during driving, there is a shift and misinterpretations follow. Therefore, a calibration check is crucial to ensure correct interpretation of the results.

Another issue which was considered important for the results was personality traits of the test person. Just as in real road conditions, extremely different reactions were observed starting from no interest for ads for the majority of the subjects to significant interest for two of them. Because of the restricted resources, 19 subjects were chosen from about 100 candidates of different age, sex, and experience as drivers. The results of three of them were not used because of problems with calibration of the eye tracking device and simulator illness. Table 1 shows sex, age and experience as driver.

Table 1. Sex, age and experience of test persons

No	Sex	Age	Experience as driver
1	F	20	2
2	F	22	1
3	F	24	6
4	F	26	8
5	F	27	8
6	F	28	10
7	F	29	10
8	M	30	12
9	F	31	14
10	M	31	13
11	M	31	13
12	M	34	15
13	M	37	15
14	M	38	20
15	M	44	27
16	F	46	30
17	F	50	33
18	M	55	36
19	M	55	39

The main interest was sensitivity to the contents of the ads. Basically, it was expected that qualitative tendencies will be found because the sample was not big and not versatile enough to formulate general opinions. To attract drivers' attention, the billboards were densely located and differently positioned. Another expectation was to identify the influence of geometric size and position in relation to trajectory of driver sight. Brightness and illumination were not tested because of the restrictions of simulator projectors. To prepare various advertising content, four independent people subjectively selected hundreds of pictures which were rather unusual and not present in real road traffic. All samples were selected by another five different unaware persons with the help of the computer display and eye-tracking system. A notable percentage of chosen advertisements had erotic content, because of the general opinion of being eye-catching and often used in outdoor advertisements. The AOI was very clearly defined as ad billboards area introduced in a city and non city driving scenario (see examples on Fig. below).

The whole time the driver's field of view included ads (3 to more than 10 big billboards of angular size more than 5° in the moment of passing). The intention was to attract drivers' attention as much as possible. The test track was about 7 km long driven for 10 - 15 min. There were 112 billboards.

Before the tests each participant had about 5 minutes to drive in a similar driving scenario but without advertisement. Drivers were not informed about the purpose of the test. They believed that they were testing eye tracking equipment.

3 Tests results

Heat maps were used as a measure of the results representing absolute gaze duration. It is a sufficient measure for preliminary analysis because it represents longer average perception and interest in the object. The value of 0.5 s was chosen as a minimum to be treated as a noticeable attraction of the driver's attention. Because the environment was specially prepared, a significantly higher number and time of fixations on advertisements was obtained compared to real road conditions. All tested persons have some fixations starting from 2-3 ads during the test drive up to over 30. Most persons look at about 10 ads. Similarly to the real road conditions, there were significant differences between tested drivers. Three of the tested persons (no 3, 10 and 16) had significantly more time and numbers of fixation than the others. It was expected that the reactions will be repeatable (e.g. the same ad and/or positioned in the same place attracts more than one driver). But it was found that it is not such a simple rule. It is interesting that different persons looked at essentially different ads but only some ads were perceived by more than one person. To give an adequate impression nearly all fixation should be presented one by one. But the valuable result of this test is that the selection of subjectively different content causes more interest than ads present in real road conditions. Erotic content was expected to be affecting male drivers. But it attracted also some females even though interest dispersion was observed for different ads among tested drivers. Evaluation of driver interest using general criteria such as erotic content or fashion or unnatural content were found not representative. There is no measure for "intensity" or oddity. To obtain qualitative statistically significant results more test persons would have to participate and/or more test drives would have to be conducted. Otherwise the attributes assigned to content would have the nature of an artefact. The main problem is that drivers have the basic task to control driving trajectory, road sign etc. Time devoted to ads was relatively short. And depending on the traffic situation and driver experience the choice of ad and time was random. In the test the environment was significantly different from that in a commercial ad test with eye-tracking. The time for perception is much longer and other tasks are less important. Nevertheless, a qualitative observation of such a simple measure as heat maps led to a significant conclusion that most of the test persons looked at very many ads which were neither the same nor comparable. Some characteristic examples of perception obtained during tests are presented in Fig. 2. to Fig. 11.

Fig. 2. presents a typical road traffic observation during driving. In Fig. 3. we can see ads competing with the traffic situation and driver behaviour was not risky.

Fig. 4. shows the same place but with a different driver and a longer distraction with another billboard with unusual content (closer to sight trajectory) but without meaningful traffic incentives.

Fig. 2. Looking at the road and oncoming car. (Eye tracking heat map registrations. Yellow colour – gaze duration longer than 0.5 s, red colour - longer than 1.0 s)

Fig. 3. A moment later. Fixations on ad with erotic content. The same driver as in Fig. 2.

Fig. 4. Another driver. The same place. Longer fixation on unusual picture. Left billboard ignored.

Fig. 5. presents a place where some fixations of different male and female drivers occurred (Registration for the same driver as in Fig. 3.).

Fig. 5. Fixation on a woman's legs wearing stockings just after the junction. The same driver as in Fig. 3.

It is difficult to define in an objective and repeatable manner the reasons why. The billboard was placed close to the eye position during driving – close to right shoulder. This is because that peripheral view can be used to control driving. It needs more investigation to find if this is content-related or due to the placement in the line of sight, contrast etc. However, other similarly placed billboards caused no fixation of any driver. The same driver ignored many other billboards on route. This means that it is really difficult to find simple rules describing such behaviour because we do not know the driver's thought processes. Another interesting result for the same driver is presented in Fig. 6. In this case the billboard of a relatively big angular size was placed centrally in the driver's field of view. A long perception occurred and the driver tried to read the descriptions on the billboard. This example might be defined as a cognitive riddle but another similar image did not attract attention. The next three examples illustrate a situation with high distraction by erotic content in a demanding traffic

 DOI: 10.1051/matecconf/201712203009

situation (female driver No.16). There were three similar billboards at the end of a T-shaped junction. In Fig. 7., Fig. 8., and Fig. 9. we can see high distraction when the driver scanned all of the billboards. Finally, a difficult traffic situation occurs (Fig. 9.) with another vehicle crossing perpendicularly.

Fig. 6. Look at a strangely looking car (upside down). Reading text. The same driver as above.

Fig. 7. Long perception of three similar billboards.

Fig. 8. A while later as above.

Fig. 9. As above. Perpendicular movement of car.

The examples above clearly show that content and placement of advertisements may influence road safety. Two other drivers looked at this ad (one of three).
A further observed phenomenon was multiplied fixations when the driver approached a billboard (Fig.10., Fig. 11.).

Fig. 10. Fixation on billboard from a distance. Other strange billboards were ignored.

Fig. 11. Multiplied fixations on the same object (Fig. 10.) while ignoring the other.

This situation confirms significant distraction, sometimes caused by trying to read the text, and can be dangerous similarly to the previous example.

4 Summary

The tests described above had a preliminary and qualitative nature and their intention was to find the possible ways and reasons why outdoor advertisements influence driver behaviour. During a typical drive there are many factors influencing drivers' perception. Distraction by advertisements depends very strongly on their content but also on sensitivity of a given driver to such particular content. The location of the billboard is also important. It was observed that if the ad is close to the central point of the driver's field of view - and if it is attractive to him/her, it is watched more often (e.g. Fig. 10.). The big number and high density of unusual billboards presented in the driving simulator environment significantly increase the time of perception in relation to real traffic for the same tested person. To obtain quantitative assessments significantly more tests would have to be conducted with longer times and more test persons. The costs of such tests may be restrictive. It may explain the inconclusive findings of other researchers [2–6,9,10,13]. The analysis of the results of above preliminary tests leads to the conclusion that for future tests the drivers should be carefully selected. Attention should be concentrated on drivers most sensitive to ads presence and content. Even if in normal traffic the percentage of such drivers is small, it is likely that they might be more exposed to accidents. The other result of the test is that drivers with more experience (e.g. professional drivers) might be less sensitive to ads as well as young drivers with less experience who are more absorbed with controlling the vehicle and observing the traffic (road signs, another vehicles). A similar situation applies to ad content which seems most important in conjunction with driver personality traits. It is not a simple task to prepare content that will attract all selected "advertisement sensitive" drivers. Nevertheless, the results of the tests show that the need is there and suggest guidelines for future tests. Using unusual ad content seems to be the right decision. The test has also confirmed that erotic content has a relatively more significant effect, however the matter is more complicated and the choice is not obvious. There was a significant dispersion of interests inside the tested group.

5 Conclusions

The highest influence on driver distraction comes from the content of advertising. Other significant factors include personality traits of the tested persons. Drivers were more commonly distracted by billboards close to the centre field of view. The illumination parameters look to be of less importance but need to be tested differently because the driving simulator has a low level of illumination and contrast. Further research is needed which should carefully preselect candidates and ad content. For this purpose carefully chosen personality tests and static computerized tests connected with eye-tracking device will be helpful. Additional factors of driver behaviour should also be measured such as speed changes, trajectory control etc. Driving scenarios can be adapted to different kinds of roads such as city and country roads or motorway. The greatest difficulty is controlling the experiment conditions. The results are highly sensitive to human behaviour, fatigue, training etc.

On the basis of the above tests no quantitative assessment is possible on the impact of ads on traffic safety. But a general observation can be very helpful with preparing future tests with a special focus on ad content. To obtain more adequate results of the influence of ads on road safety, testing conditions must be fixed. As the test showed some people (three of nearly 20) look at a really big number of ads.

Because tests with real drivers are costly and time consuming, a careful preparation of scenarios and selection of test persons is important for reliable findings for a moderate price. There are plans to conduct a more detailed analysis of fixations and saccades of registered data which might give some more information. But at this stage is it clear that it is difficult to find general rules or universal examples of ad content which may be used for all tested persons. Therefore more attention should be paid to a proper choice of test drivers who are "ads sensitive". Such an experiment cannot be repeated too many times because of the effect of habit.

Acknowledgements

Project "The impact of advertising on the level of traffic safety" is financed by the National Centre for Research and Development (NCBiR) and General Directorate for National Roads and Motorways (GDDKiA), grant agreement DZP/RID-I-33/4/NCBR/2016

References

1. M.J. Daluge, M. DeLong, L. Hanig, Outdoor Advertising Control Practices in Australia, Europe, and Japan. International Technology Scanning Program Report FHWA-PL-10-031 (Federal Highway Administration U.S. Department of Transportation, American Association of State Highway and Transportation Officials, 2011)
2. T. Dukic, C. Ahlstrom, C. Patten, C. Kettwich, K. Kircher, Effects of Electronic Billboards on Driver Distraction, Traffic Inj. Prev. **14(5)** pp. 469–476 (2013). doi:10.1080/15389588.2012.731546
3. S.E. Lee, M.J. Mcelheny, R. Gibbons, Driving Performance and Digital Billboards, Prepared for: Foundation for Outdoor Advertising Research and Education (2007)
4. B. Metz, H.-P. Krüger, Do supplementary signs distract the driver?, Transp. Res. Part F Traffic Psychol. Behav. **23** pp. 1–14 (2014). doi:10.1016/j.trf.2013.12.012
5. A. Smiley, B. Persaud, G. Bahar, C. Mollett, C. Lyon, T. Smahel, W. Kelman, Traffic Safety Evaluation of Video Advertising Signs, Transp. Res. Rec. J. Transp. Res. Board. **1937** pp. 105–112 (2005). doi:10.3141/1937-15

6. B. Wallace, Driver distraction by advertising: genuine risk or urban myth?, in: Proc. Inst. Civ. Eng., (2003)
7. K. Wandachowicz, M. Zalesińska, K. Domke, S. Mroczkowska, P. Skrzypczak, Wielkopowierzchniowe reklamy z diodami świecącymi a bezpieczeństwo ruchu drogowego, Przegląd Elektrotechniczny. **87** pp. 73–77 (2011)
8. S. Bendak, K. Al-Saleh, The role of roadside advertising signs in distracting drivers, Int. J. Ind. Ergon. **40** pp. 233–236 (2010). doi:10.1016/j.ergon.2009.12.001
9. M.S. Young, J.M. Mahfoud, N.A. Stanton, P.M. Salmon, D.P. Jenkins, G.H. Walker, Conflicts of interest: The implications of roadside advertising for driver attention, Transp. Res. Part F Traffic Psychol. Behav. **12** pp. 381–388 (2009). doi:10.1016/j.trf.2009.05.004
10. D. Kaber, C. Pankok, B. Corbett, W. Ma, J. Hummer, W. Rasdorf, Driver behavior in use of guide and logo signs under distraction and complex roadway conditions, Appl. Ergon. **47** pp. 99–106 (2015). doi:10.1016/j.apergo.2014.09.005
11. S. Milloy, J. Caird, External Driver Distractions:The Effects of Video Billboards and Wind Farms on Driving Performance, in: Handb. Driv. Simul. Eng. Med. Psychol., (CRC Press, 2011). doi:10.1201/b10836-17
12. F. Schieber, K. Limrick, R. McCall, A. Beck, Evaluation of the Visual Demands of Digital Billboards Using a Hybrid Driving Simulator, Proc. Hum. Factors Ergon. Soc. Annu. Meet. **58** pp. 2214–2218 (2014). doi:10.1177/1541931214581465
13. A. Megías, L.L. Di Stasi, A. Maldonado, A. Catena, A. Cándido, Emotion-laden stimuli influence our reactions to traffic lights, Transp. Res. Part F Traffic Psychol. Behav. **22** pp. 96–103 (2014). doi:10.1016/j.trf.2013.09.017

Roadside advertising and the distraction of driver's attention

Adam Tarnowski [1,*], *Anna* Olejniczak-Serowiec[2], and *Agnieszka* Marszalec[2]

[1]Motor Transport Institute, Road Traffic Safety Centre, Warsaw, Poland
[2]Univeristy of Warsaw, Faculty of Psychology, Warsaw, Poland

Abstract. Distraction during driving is becoming a major problem in contemporary transport and traffic psychology. Concentration may deteriorate complex vehicle systems due to the provision of unnecessary information and use of mobile phones (the problem is not only talking but writing text messages and e-mails, browsing sites, etc.). A significant role is also played by advertisers who use aggressive ways to attract attention and communicate product information, especially because they compete with an already overloaded attention system. On the other hand, the need for stimulation is strong with people increasingly less tolerant to monotony. The RoAdvert project is aimed to develop evidence-based rules of placing roadside advertising with respect to safety and real possibilities of regulating the advertising market, including the optimal level of driver stimulation. The paper will present a preliminary analysis of the survey and experimental research.

1 Introduction

It is generally acknowledged, that attention focus and situational awareness are essential for vehicle driving [1,2], and that attention distraction and overload are key reasons for missing important on-the-road events [3,4]. Such errors, called inattentional blindness, pose a serious threat to driving safety [5], with the risk of a crash increasing when the individual's attention capacity decreases[6]. In order to maintain situational awareness, the driver needs to employ a number of cognitive mechanisms: perception, comprehension, memory (i.e. working memory) and anticipation [7,8]. The latter, which actually assumes that the situation changes over time, is based on proper time perception and is thus dependent on attention and cognitive load[9]. Apart from tracking multiple moving objects, the second attention function crucial for driving is detecting changing items in the environment [10]. Research data prove, that attention distraction leads not only to a lack of conscious detection of changes in one's environment, but also to the unconscious perception failure[11]. This is especially true, for peripheral change detection (e.g. a pedestrian or a cyclist entering the car path) while performing attention-demanding primary tasks (i.e. car driving) [12].

Statistical analysis of the number of accidents and behaviour changes reveals some decrease in driver effectiveness in the presence of roadside advertisement [13] which is usually attributed to distraction [14–17] because advertisements compete for drivers' attention. Also, drivers declare higher workloads in the presence of roadside advertising [15,16,18]. Research focused on advertisement content, on the other hand, reveals a link

* Corresponding author: adam.tarnowski@its.waw.pl

between the advertisement's sexual vs. nonsexual appeal and cognitive functioning of people to whom they were presented. It seems that sexual advertisements are not only subjectively assessed and get more attention [19], but they are also better memorized [20,21]. Nonetheless, the performance of cognitive tasks decreases under the influence of sexual advertisements [19], even when memorizing brand-related information included in the advertisement itself is concerned [21].
The above literature review raises a question whether the presence of roadside advertising causes visual attention distraction, increases driver cognitive workload and diverts working memory from the main task, and - if so - whether the drivers' individual attention capacity and advertisement content moderate the effects. The two studies described below were conducted to answer these questions.

2 Study I

Study I is a reanalysis of the data gathered during EYEVID project on road infrastructure (PBS1/B6/9/2012). The project developed new evidence based tools for safety inspection of existing roads. The reanalysis is designed to verify the impact of roadside advertising on driver attention. We investigated whether visual noise decreases the amount of attentional resources engaged into a primary task (i.e. car driving), and the influence of this factor on visual system overload.

2.1 Hypotheses

It was hypothesized that:
1. In the presence of a roadside advertisement, visual attention distraction increases.
2. Roadside advertisement increases cognitive load while driving.
3. The impact of roadside advertisement on drivers' focus and cognitive load depends on their attention capabilities
   a. Drivers with high attention performance are less affected by the roadside advertisement than those with poor attention performance.

2.2 Participants and procedure

The data come from 45 active drivers (14 female and 31 male) with a minimal experience of 10 years. Participants' age varied between 28 and 62 (M=42.67; SD=11.91). All the participants took part in a 6 kilometre simulated drive and attention test.

Attention was measured with a Pop-Up Test (part of the Test2Drive diagnostic system) [22]. The test is designed to assess two attention functions: searching and identifying crucial objects and lasts about 3 minutes.

The simulated driving was performed in an Opel Astra based AS1200-6 simulator. In the simulator a faithful reconstruction of a 6 kilometre route was programmed. The route consisted of rural and urban roads with high vs. low advertising density. Participants were instructed to drive the car as they usually do and search for service stations. The procedure was as follows: training drive, attention test, test drive.

During the drive, participants' eyes were followed with the use of SMI Glasses 30 Hz. The area of interest for task performance were: road, mirrors, and cockpit. The percentage of time when eyesight was engaged in the areas of interest was counted as a visual attention focus measure, and the average pupil size was used as a cognitive load measure instead of self-description [23,24].

2.3 Results

The results show that the density of advertising affects the percentage of time in which participants' sight was engaged in the areas of interest. The difference between the high advertising density and low advertising density was statistically significant ($F=23.06$; $p<0.001$; $\eta2=0.385$). For the pupil size no statistically significant effects were observed ($F=0.346$; $p>0.05$; $\eta2=0.013$). The interaction effects between poor and high attention test performance was insignificant both for dwell time proportion ($F=0.49$; $p>0.05$, $\eta2=0.017$), and for the average pupil size ($F=1.86$; $p>0.05$, $\eta2=0.067$). The results for dwell time proportion are illustrated in Figure 1 and the results for the average pupil size are illustrated in Figure 2.

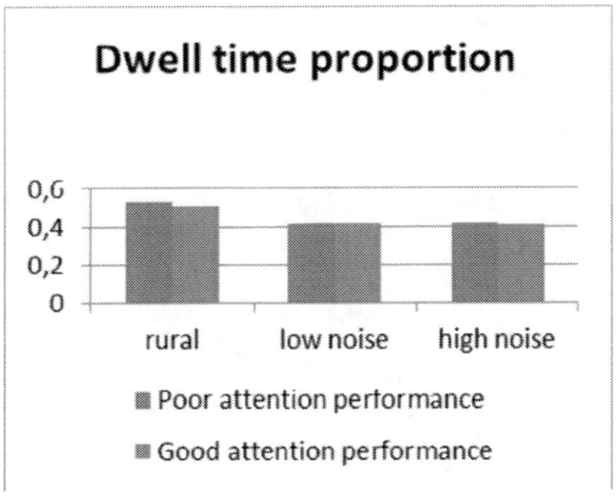

Fig. 1. Dwell time proportion in different conditions divided by attention test performance.

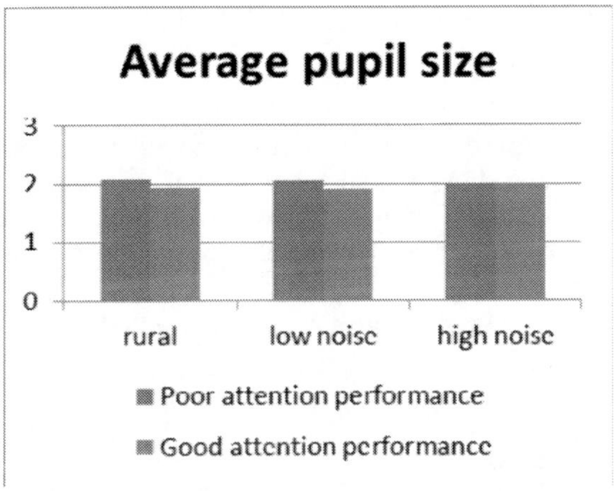

Fig. 2. Average pupil size in different conditions divided by attention test performance.

2.4 Conclusions

On the basis of the data presented, it can be concluded that roadside advertising decreases visual attention focus (as stated in hypothesis 1), however it does not necessarily affect cognitive load (contrary to hypothesis 2). No evidence supporting hypothesis 3 was found.

3 Study II

The study's aim was to verify the impact of advertisement contents on memory [25].

3.1 Hypotheses

It was hypothesized that:

1. In the presence of roadside advertisements, drivers' working memory capacity decreases.
2. The impact of roadside advertisement on drivers' working memory capacity depends on the advertisement content
   a. the presence of human representation in the advertisement decreases working memory performance
   b. working memory performance is worse in the context of sexual advertisement than in the nonsexual.

3.2 Participants and procedure

The experiment was conducted on 31 (15 female and 16 male) Polish amateur drivers aged from 19 to 45 (M=26.6, SD=1.3). Their average length of driving experience was 7.5 year (SD=7.23).

The stimuli consisted of 38 photos of driving scenes (2 training photos and 36 test photos), some of them including an advertising billboard. In the advertisement-present scenes three conditions were specified: advertisement without human representations vs. with human representation, the latter was then divided into sexual (moderately sexual contents – people wearing underwear) and nonsexual contents. The stimuli was presented for 2 seconds, subsequently, the participant was asked which traffic signs were present in the photo. Sex of the person presented in the advertisement was controlled. The number of errors in the memory task was measured.

3.3 Results

The memory test results differed significantly between advertisement vs. no-advertisement condition. Further analyses show that there is also a statistically significant difference between nonhuman and human condition. Almost no differences occurred between sexual and nonsexual contents condition. The tests results are presented in Table 1 and Figure 3.

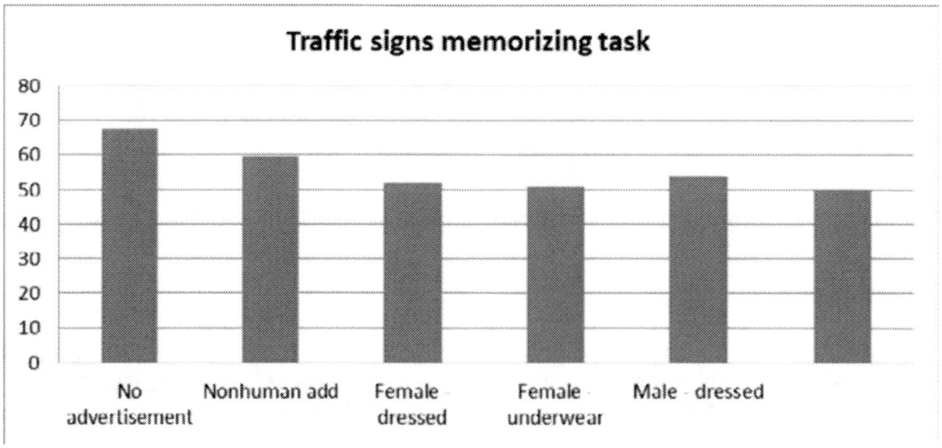

Fig. 3. The presence of advertisement and working memory performance

Table 1. T-test for dependent samples results - differences between different types of advertising.

Conditions pair	T test	Statistical significance	Effect size
No advertisement - Nonhuman add	t=-3.80	p=0.001	d=-0.683
No advertisement - Male underwear	t=-10.84	p=0.001	d=-2.067
No advertisement - Male dressed	t=-9.52	p=0.001	d=-1.711
No advertisement - Female underwear	t=-9.78	p=0.001	d=-1.818
No advertisement - Female dressed	t=-10.04	p=0.001	d=-1.947
Female underwear - Male underwear	t=0.77	p=0.45	d=0.138
Female dressed - Male dressed	t=-0.99	p=0.33	d=-0.191
Male underwear - Nonhuman add	t=-7.44	p=0.001	d=1.428
Male dressed - Nonhuman add	t=-3.09	p=0.004	d=0.556
Female underwear - Nonhuman add	t=-5.27	p=0.001	d=-0.966
Female dressed - Nonhuman add	t=-4.98	p=0.001	d=-0.936
Male dressed - Male underwear	t=-2.07	p=0.047	d=-0.392
Female dressed - Female underwear	t=-0.65	p=0.518	d=0.118
Female underwear - Male dressed	t=-1.344	p=0.189	d=-0.247
Female dressed - Male underwear	t=1.768	p=0.087	d=0.317

3.4 Conclusion

The research confirms prior knowledge about roadside advertisement as a potential attention and perceptiveness distractor (hypothesis 1 confirmed). Moreover, the aforementioned results form a basis to conclude that advertisement content does affect driver attention (i.e. working memory), namely, the presence of human representation increases distraction. Moderate sexual context, however, does not reinforce the effect (hypothesis 2 partly confirmed).

4 General Discussion

Consistently with prior research, drivers' visual distraction in the context of roadside advertisements was confirmed in the present studies. Their perceptiveness and memorizing – key processes for situational awareness and safe driving – decreased when the roadside advertisement was present. Additionally, some influence of the advertisements' content was observed, which was an increase in distraction when the advertisement included human representation. However, the effect was indifferent to the presence of sexual context. Contrary to the expectations, no moderation effect of attention capacity was observed. This effect, however, may be due to the fact, that crash risk and attention capacity correlation observed in older drivers [2] occurs only until a certain level of attention capacity and when the level is reached, further attention capacity increase does not affect driving effectiveness.

Acknowledgements

The research and the preparation of this article were founded by the National Centre for Research and Development and the General Director for National Roads and Motorways (DZP/RID/I-33/4/NCBR/2016).

References

1. L.J. Gugerty, Situation Awareness During Driving: Explicit and Implicit Knowledge in Dynamic Spatial Memory, J. Exp. Psychol. Appl. **3** pp. 42–66 (1997)
2. A. Cuenen, E.M.M. Jongen, T. Brijs, K. Brijs, M. Lutin, K. Van Vlierden, G. Wets, Does attention capacity moderate the effect of driver distraction in older drivers?, Accid. Anal. Prev. **77** pp. 12–20 (2015). doi:10.1016/j.aap.2015.01.011
3. I.E. Hyman, S.M. Boss, B.M. Wise, K.E. McKenzie, J.M. Caggiano, Did you see the unicycling clown? Inattentional blindness while walking and talking on a cell phone, Appl. Cogn. Psychol. **24** pp. 597–607 (2009). doi:10.1002/acp.1638
4. A. Stinchcombe, S. Gagnon, Driving in dangerous territory: Complexity and road-characteristics influence attentional demand, Transp. Res. Part F Traffic Psychol. Behav. **13** pp. 388–396 (2010). doi:10.1016/j.trf.2010.06.008
5. I.D. Brown, Review of the "looked but failed to see" accident causation factor, in: Behav. Res. Road Saf., pp. 116–124 (London, 2002)
6. K.R. Johannsdottir, C.M. Herdman, The Role of Working Memory in Supporting Drivers? Situation Awareness for Surrounding Traffic, Hum. Factors J. Hum. Factors Ergon. Soc. **52** pp. 663–673 (2010). doi:10.1177/0018720810385427
7. M.R. Endsley, Situation Awareness Misconceptions and Misunderstandings, J. Cogn. Eng. Decis. Mak. **9** pp. 4–32 (2015). doi:10.1177/1555343415572631
8. R.A. Block, D. Zakay, Models of psychological time revisited, in: Time Mind, pp. 171–195 (Hogrefe & Huber, 1996)
9. L.M. Trick *, J.T. Enns, J. Mills, J. Vavrik, Paying attention behind the wheel: a framework for studying the role of attention in driving, Theor. Issues Ergon. Sci. **5** pp. 385–424 (2004). doi:10.1080/1463922041233129893
10. A. Mack, I. Rock, Inattentional Blindness (MIT Press, Cambridge, 1998)
11. D.J. Simons, C.F. Chabris, Gorillas in Our Midst: Sustained Inattentional Blindness for Dynamic Events, Perception. **28** pp. 1059–1074 (1999). doi:10.1068/p281059
12. G. Yannis, E. Papadimitriou, P. Papantoniou, C. Voulgari, A statistical analysis of the impact of advertising signs on road safety, Int. J. Inj. Contr. Saf. Promot. **20** pp. 111–120 (2013). doi:10.1080/17457300.2012.686042
13. A. Smiley, B. Persaud, G. Bahar, C. Mollett, C. Lyon, T. Smahel, W. Kelman, Traffic Safety Evaluation of Video Advertising Signs, Transp. Res. Rec. J. Transp. Res. Board. **1937** pp. 105–112 (2005). doi:10.3141/1937-15
14. S. Bendak, K. Al-Saleh, The role of roadside advertising signs in distracting drivers, Int. J. Ind. Ergon. **40** pp. 233–236 (2010). doi:10.1016/j.ergon.2009.12.001
15. M.S. Young, J.M. Mahfoud, N.A. Stanton, P.M. Salmon, D.P. Jenkins, G.H. Walker, Conflicts of interest: The implications of roadside advertising for driver attention,

Transp. Res. Part F Traffic Psychol. Behav. **12** pp. 381–388 (2009). doi:10.1016/j.trf.2009.05.004

16. T. Horberry, J. Anderson, M.A. Regan, T.J. Triggs, J. Brown, Driver distraction: The effects of concurrent in-vehicle tasks, road environment complexity and age on driving performance, Accid. Anal. Prev. **38** pp. 185–191 (2006). doi:10.1016/j.aap.2005.09.007
17. D. Crundall, E. Van Loon, G. Underwood, Attraction and distraction of attention with roadside advertisements, Accid. Anal. Prev. **38** pp. 671–677 (2006). doi:10.1016/j.aap.2005.12.012
18. J. Edquist, T. Horberry, S. Hosking, I. Johnston, Effects of advertising billboards during simulated driving, Appl. Ergon. **42** pp. 619–626 (2011). doi:10.1016/j.apergo.2010.08.013
19. T. Reichert, S.E. Heckler, S. Jackson, The Effects of Sexual Social Marketing Appeals on Cognitive Processing and Persuasion, J. Advert. **30** pp. 13–27 (n.d.). doi:10.2307/4189167
20. J. King, A. McClelland, A. Furnham, Sex Really Does Sell: The Recall of Sexual and Non-sexual Television Advertisements in Sexual and Non-sexual Programmes, Appl. Cogn. Psychol. **29** pp. 210–216 (2015). doi:10.1002/acp.3095
21. L. Samson, The Effectiveness of Using Sexual Appeals in Advertising, J. Media Psychol. pp. 1–12 (2016). doi:10.1027/1864-1105/a000194
22. A. Tarnowski, TEST2DRIVE Podręcznik użytkownika (ALTA, Siemianowice Śląskie, 2016)
23. D. Kahneman, Attention and effort. (Prentice-Hall, New Jersey, 1973)
24. G. Porter, T. Troscianko, I.D. Gilchrist, Effort during visual search and counting: Insights from pupillometry, Q. J. Exp. Psychol. **60** pp. 211–229 (2007). doi:10.1080/17470210600673818
25. A. Marszalec, Unpublished master thesis: Driver attention in relation to roadside advertisement visual content, University of Warsaw, 2016

Modelling fatalities on regional road networks

Joanna Wachnicka[1,*]

[1]Gdansk University of Technology, Faculty of Civil and Environmental Engineering, Gdansk, Poland

Abstract. During the last decade Poland's road fatalities went down by 44%. The trend differs from region to region. Effective road safety management in regions requires tools for forecasting road safety measures and identifying factors influencing road fatality numbers. Mathematical models can provide such tools. They take into account local characteristics such as: demography, economy, infrastructure and motorization. Such models could be used for better regional road safety management. This paper presents an attempt to build such models used for forecasting road fatalities in EU regions and Polish voivodships.

1 Introduction

Road accidents worldwide claim nearly 1.3 m lives annually causing an economic loss that represents 2 - 3% of the world's GDP. Sadly Poland ranks among the top thirty countries in the world and the European Union's top five for its road traffic risk. Despite the EU being the world leader on road safety, four countries: Germany, France, Italy and Poland contribute to more than 50% of the EU's road fatalities. The risk of becoming a fatality in relation to Poland's population is double the European Union's average and as much as three times more than in the United Kingdom, the Netherlands and Sweden.

It is estimated that until 2030 road fatalities may double if the worldwide community fails to take effective and comprehensive steps to improve road safety. Being involved in a road accident may become one of the five most frequent causes of death of the globe's population. This is what spurs a variety of government organisations and NGOs into action geared towards reducing road accidents. The United Nations with its World Health Organisation (WHO) have prepared recommendations for the countries, designed to improve road safety. They are part of the Decade of Action for Road Safety adopted in 2011 (for the years 2011 – 2020). The fundamental intention behind this idea is to save 5 million people from death in a road accident in the years 2011 – 2020. This is to be achieved by efforts delivered by the particular countries, including the development and implementation of road safety programmes, appointing road safety management bodies, effective accident data collection and increasing spending on road infrastructure to ensure that all road projects comply with safety requirements when they are designed, built and maintained [1].

Poland has its own systemic solutions designed to improve road safety. The country's first road safety programme (adopted by the government) was developed in 1972 [2]. In 1993 the National Road Safety Council was established at the central level with Regional Road Safety Councils appointed at the regional level. A year later the minister of transport asked the Committee of Scientific Research to conduct a research project called "Integrated Road Safety Programme" under the acronym of GAMBIT. Developed by an interdisciplinary team

* Corresponding author: joanna.wachnicka@wilis.pg.gda.pl

of experts headed by professor R. Krystek, the programme set quantitative targets to be attained within the next 5 years and the projects to be implemented within this time. Two years into the programme it was established that the targets cannot be reached. More GAMBIT road safety programmes followed in 2000, 2005 and in 2012. While they were nation-wide programmes, new programmes were designed specifically for the regions and cities to ensure that action at the lower levels can have a stronger effect on the regional problems. The regional or city road safety forecasts which they used were largely based on police statistics and earlier domestic and international experience. What they lacked, however, were mathematical tools to support road safety forecasts such as mathematical models that take account of regional characteristics and their effect on the regional situation. Because no country has solved the problem of modelling regional road safety, the decisions were made to build such road safety models for regions in the United States, Europe and Poland. Such models, due to the inclusion of multiple factors, may be helpful in analysing the impact of the individual on the modelled variable, thus helping to better manage road safety in the regions.

2 Characteristics of regional road safety

Road safety on the US road network was analysed using data from 51 states. In the case of EU regions, data were collected from 188 NUTS 2 regions of the following countries: the United Kingdom, Czech Republic, Germany, Poland, France, Spain, Slovakia, Hungary, Sweden, Bulgaria and Romania. In Poland, the analysis looked at 16 regions (voivodeships) (Fig. 1). The time period under analysis in all the cases spanned from 1999 to 2008 with Polish regions as the exception because the data covered the years 1999-2014. Detailed analyses of the levels of road safety in groups of regions showed significant differences between them.

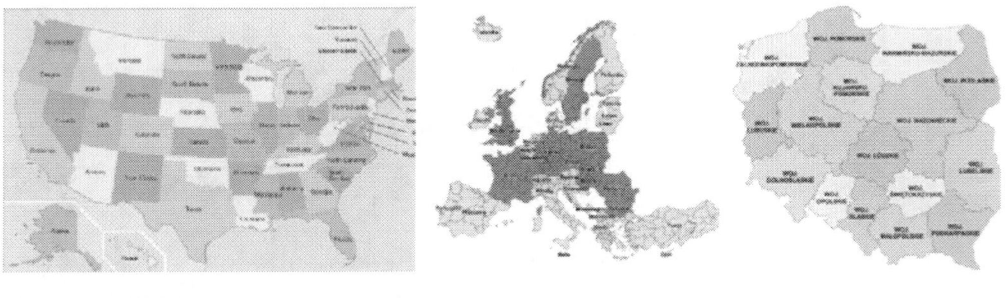

Fig. 1. Analysed regions in particular data groups.

US regions showed strong regional differentiations. As an example, the population in the states ranges from 480,000 to more than 36 m people and the area is between 157,000 km^2 to app. 1.5 m km^2. Population density shows strong variations as well with values ranging from 420 people/km^2 to more than 5,500 people/km^2. Road accident fatalities show a range between 34 to app. 4,300 people annually and about 800 people on average annually. The population of the EU regions studied is between 70,000 to 11.7 m and the area is from 100 km^2 to 94,000 km^2. Population density varies from 5 people/km^2 to more than 5,500 people/km^2. Road accident fatalities are within a range of 1 to almost 1,150 people, with the average of 213 people. The area of Polish voivodeships ranges from app. 9.4 to 35,600 km^2, population from app. 1 do 5.3 m, and the number of road fatalities regionally varies from 81 to more than 1,100 people annually.

The difference in the average area of regions in the United States, Europe and Poland is shown in Fig. 2. The figure also presents the relation between fatalities and the region's population. As we can see, European and Polish regions are similar while the American states differ significantly from the two groups. The average gross domestic product GDPPC in the US is higher than Europe's by 25% and almost three times higher than a Polish voivodeship's GDPPC. At the same time the average fatalities on the roads of the particular US states are almost four times higher than in Europe and twice as high as that in Poland.

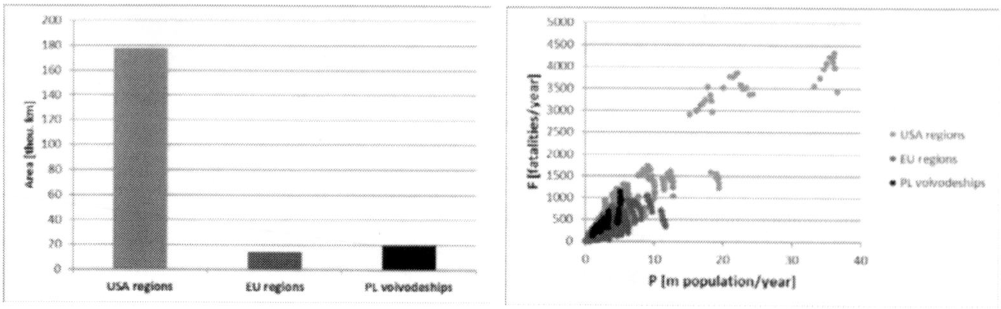

Fig. 2. Analysed regions in particular data groups.

Table 1. Average, minimal and maximal values for the basic characteristics in the three analysed databases.

Data base	Characteristics	average	minimum	maximum
States AP	Area [thou.km^2]	177.572	0.157	1464.195
	Population [m]	5.712	0.480	36.580
	Motorization rate [veh./person]	0.862	0.253	1.512
	Gross Domestic Product [thou. ID/person]	41.200	25.039	151.392
	Fatalities [people]	812	34	4329
	Fatality rate [people/100 thou.mk.]	16.134	1.032	39.408
Regions EU	Area [thou.km^2]	14.486	0.32	154.312
	Population [m]	2.081	0.119	11.694
	Motorization rate [veh./person]	0.535	0.098	1.603
	Gross Domestic Product [thou. ID/person]	30.183	5.686	116.83
	Fatalities [people]	162	6	1057
	Fatality rate [people/100 thou.mk.]	9.861	1.055	25.839
Voivodeships PL	Area [thou.km^2]	19.524	9.412	6.645
	Population [m]	2.396	1.008	5.204
	Motorization rate [veh./person]	0.357	0.166	0.588
	Gross Domestic Product [thou. ID/person]	12.069	7.932	25.951
	Fatalities [people]	367	128	1146
	Fatality rate [people/100 thou.mk.]	15.674	9.145	24.554

Preliminary analyses have shown that regional data cannot be analysed in one set. This is because the scale of the regions differs significantly, as does variable availability and the scope of change in the analysed risk measures. As a consequence, the analyses that followed were conducted separately for each group of regions.

A number of independent variables were collected to build the mathematical models. Divided into groups of factors, the variables are region-specific. The following are the groups of factors:
- demographic (characteristic of the region's population),
- geographic (characteristic of the area and land use),

- motorization (characteristic of the fleet of vehicles using the region's road network),
- mobility (characteristic of the mobility of the population and their transport behaviour),
- infrastructure (characteristic of the structure of the region's road network),
- economic (characteristic of the region's economic development and its investment spending),
- social (characteristic of the region's population in terms of its development, health, behaviour).

Data availability differed from group to group and amounted to 49 independent variables in US regions, 27 in European regions and 71 in Polish regions. While preliminary analyses of the factors are presented in the reviewed articles [3,4], the focus of this article is to describe the models developed for each of the three groups of regions.

3 Modelling methods

Preliminary data analyses [3,5] made it clear that modelling fatality numbers was going to be difficult. As a result, the decision was made to create models of the relative fatality rate, the *RFR*. Where:

$$RFR_i = \frac{F_i}{P_i} \quad (1)$$

where:
- RFR_i – fatality rate in the region i [fatalities /100 thou. people/year]
- F_i –fatalities in the region i [fatalities /year]
- P_i – population of the region i [fatalities /100 thou. people/year].

This approach helps to build reliable models and, using the region's forecasted population, calculate the forecasted number of fatalities in relation to specific factors using this formula:

$$F_i = RFR_i \cdot P_i \quad (2)$$

To understand which independent variables may have an effect on the RFR, for each database an analysis was made into the most probable functions which could describe the relations under study.

The following functions were tested: linear, logarithmic, Kuznets, exponential, power and power-exponential. As a result, it was possible to identify the link functions in the new models. The exponential and power-exponential functions were found to work best in the case.

4 Results of modelling

In the case of states in the US, the relative fatality rate was strongly influenced by: length of vehicle kilometres travelled per one inhabitant *VKTPC* and income per capita in that area *INPC*. Thanks to variable availability, the models reached a quality factor of app. 75%. Formula 3 shows an example of one of the models.

$$RFR = 14.79 \cdot exp\left(-0.03 \cdot INPC + 0.04 \cdot DPR_{(D)} + 0.09 \cdot VKTPC + 0.11 \cdot DME_{(D)}\right) \quad (3)$$

where:
- $INPC$ – national income per capita [thou.ID/person],
- $DPR_{(D)}$ – demographic density of paved roads [thou.km/100 thou. person],
- $VKTPC$ – kilometres travelled by vehicles per capita [thou.km/person],

- $DME_{(D)}$ – demographic density of motorways and express roads [thou.km/100 thou. person].

Analysis of coefficients for the particular variables shows that as national income per capita *INPC* grows, the relative fatality rate on the state's roads goes down. At the same time, an increase in vehicle kilometres travelled per capita, demographic density of paved roads and demographic density of motorways and express roads increases the risk of becoming a road accident fatality.

When the *RFR* was modelled for European regions, power and exponential functions were used. The level of socio-economic development expressed as gross domestic product per capita *GDPPC* turned out to be positively correlated. In the case of vehicle density *DV*, it was observed that as the rate begins to grow, so does the relative fatality rate and once it reaches maximum it begins to decrease.

$$RFR = 13.44 \cdot DV^{-0,84} \cdot exp(0.01 \cdot GDPPC - 0.03 \cdot DR_{(A)} + 2.28 \cdot DV) \qquad (4)$$

where:
- *DV* – total vehicle density [veh./km^2/year],
- *GDPPC* – gross domestic product per capita [thou.ID/person/year],
- $DR_{(A)}$ – total road density [km/km^2/year].

Unfortunately, none of the European regional models exceeded a quality factor of 45%, which is probably due to how strongly differentiated the regions are. Another possible reason may be the short time interval of the analysis. The problem requires further analyses as the work continues.

Because Poland is struggling with its significant fatality rates, the next stage of the work focussed on models of the relative fatality rate in Polish regions. Thanks to detailed data availability, the models use a bigger number of explanatory variables. The resulting model achieved a quality factor of R^2=60%. In the case of Polish data, detailed preliminary analyses showed that being a capital region matters. In addition, the relative fatality rate in Polish voivodeships decreases as gross domestic product per capita and total vehicle density increase. Regional road safety improves when regional roads are modernised and more people live in cities.

$$RFR = 24,07 \cdot S^{0,23} \cdot GDPPC^{-0,42} \cdot exp(-1,02 \cdot DV + 0,33 \cdot DR_{(D)} + 0,51 \cdot PUP - 0,75 \cdot REXV + 0,04 \cdot PUA - 0,07 \cdot PUP + 29,27 \cdot DME_{(D)}) \qquad (5)$$

where:
- *GDPPC* – gross domestic product per capita [thou.ID/person/year],
- *S* – variable denoting the capital voivodeship [capital=10, others=1],
- *DV* – total vehicle density [veh./km^2/year],
- $DR_{(D)}$ – total demographic road density [thou.km/100 thou. person/year],
- *PUP* – share of urban population [%],
- *REXV* – spending on regional roads [m PLN/km/year],
- *PUA* – share of built-up and urbanised land in overall area [%],
- $DME_{(D)}$ – demographic density of motorways and express roads [thou. km/100 thou. person/year].

5 Discussion of the results

The models' quality factors varied, depending on data availability and the scope of the regions being modelled. A comparison of the results shows that it is important to collect and share data about the number of kilometres [6,7] travelled on the region's road network. This factor is critical for understanding risk exposure. Collecting data about road expenditure also turned out to be important [8,9]. This may show indirectly what changes are made to the road network to improve road safety. The results of the modelling confirmed the conclusions from previous research which looked at the effect of regional economic growth on road safety [10,11]. The results of the modelling using three groups of regions suggest that it is difficult to build models to give sufficient explanation of the relations, if the regions are highly differentiated and detailed variables are not available.

6 Conclusions

There are many reasons why modelling fatalities is a difficult task. First, if the aim is to build mathematical models that can sufficiently explain the relations, the dependent variable must be as compact as possible. When groups of regions vary significantly in terms of their size, population and, often as a result, fatalities, it is advisable to normalise the data by introducing a relative variable just as was the case in this article. Second, it is important to collect independent variables carefully and with some differentiation to obtain a description of the region's mobility, infrastructure, economy and social aspects. It is also advisable to have expert knowledge of the region to ensure that qualitative variables can be added during modelling to differentiate it from other regions.

References

1. L. Michalski, K. Jamroz, D. Gajda, Wstępna ewaluacja interim Krajowego Programu Bezpieczeństwa Ruchu Drogowego GAMBIT 2005, in: IX Int. Road Saf. Semin. GAMBIT 2012, pp. 20–21 (Gdansk, 2012)
2. Resolution of the Council of Ministers No. 155/72 of 9 June 1972 on directions and measures intended to improve road safety (Poland, 1972)
3. J. Wachnicka, Modelling selected road safety measures at the regional level in Europe, J. Polish Saf. Reliab. Assoc. **3** pp. 251–258 (2012)
4. J. Wachnicka, K. Jamroz, Identification and analysis of factors influencing road safety in regions, in: Road Saf. Simul. Int. Conf. RSS2013, (Rome, 2013)
5. J. Wachnicka, Identification and comparative analysis of factors influencing road safety in US regions and in Polish voivodeships, Transp. Probl. **8** pp. 53–66 (2013)
6. L.S. Robertson, Reducing death on the road: the effects of minimum safety standards, publicized crash tests, seat belts, and alcohol., Am. J. Public Health. **86** pp. 31–34 (1996)
7. G. Scott, G. Pittams, N. Derby, Regression models of New Zealand road casualty data: results of a preliminary investigation, in: Australas. Transp. Res. Forum, pp. 829–854 (Wellington, 1987)
8. S. Gaca, W. Suchorzewski, M. Tracz, *Inżynieria ruchu drogowego. Teoria i praktyka (Road traffic engineering. Theory and practice)* (WKŁ, Warsaw, 2008)
9. J. Kempa, Najczęstsze wady infrastruktury drogowej, in: Bezpieczeństwo Ruchu Miej., pp. 239–287 (WKŁ, Warszawa, 2005)
10. J.Y. Douglas, T.W. Likens, Alcohol Regulation and Auto Fatalities, Int. Rev. Law Econ. **20** pp. 107–126 (2000). doi:10.1016/S0144-8188(00)00023-5
11. S. Hakim, et al., A Critical Review of Macro Models for Road Accidents, Accid. Anal. Prev. **23** pp. 379–400 (1991)

New innovative educational method to prevent accidents involving young road users (aged 15-24) – European Road Safety Tunes

Dagmara Jankowska-Karpa [1,*], and *Justyna* Wacowska-Ślęzak [1]

[1]Motor Transport Institute, Road Safety Centre, Warsaw, Poland

Abstract. The article presents a new teaching method designed to improve road safety among young road users. Developed under "European Road Safety Tunes", this international project was co-funded by EU DG MOVE. Its main aim is to improve road safety and minimize the number of road accidents, injuries and fatalities among road users who are 15-24 years old. The Safety Tunes method contains a series of workshops addressed to young vocational school students: cyclists, moped and motor riders and car drivers. The workshops incorporate peer and emotive education, and delivery of road safety related messages through different types of artistic forms. The topics tackled during class address awareness of possible risks and risk-behaviour, prevention of distraction and reduction in young fatalities and serious injuries on the road. All actions within the project are evaluated, both in terms of the impact of the workshops on students' attitudes towards road safety problems and in terms of process assessment.

1 About the European Road Safety Tunes project

The project European Road Safety Tunes (short: Safety Tunes) is delivered by an international consortium composed of eight Partner countries: Austria, Belgium, Czech Republic, Hungary, the Netherlands, Poland, Slovenia and Spain. The project is co-financed by EU DG MOVE and refers to EU policy orientations on road safety 2011-2020 from the EC 2020 strategic objective N7 to protect vulnerable road users. Each year more than 4,000 young people aged 15-24 are killed in road accidents across Europe. They are the reason why Safety Tunes is focusing on the target group of young road users – cyclists, moped-riders, motorcyclists and car-drivers [1].

Safety Tunes uses emotive and peer-oriented measures to teach road safety topics in vocational schools and addresses:

- awareness of possible risks and risk-behaviour of the general prospective driving population concerning vulnerable road users,
- prevention of distraction,
- reduction in young drivers' fatalities and serious injuries on the road.

There is a cross connection between young people with vocational level education, unsafe driving behaviour and a higher accident rate. During the past years European secondary schools have introduced road accident prevention topics. Moreover, risk management is often taught in a cognitive way by showing facts and figures. However, research suggests that a

* Corresponding author: dagmara.jankowska@its.waw.pl

more emotive transfer of social know-how is by far more effective. Emotional and creative methods can trigger a feeling of happiness and hence the learned experiences are more memorable [2].

Safety Tunes methodology intends to transfer road safety know-how through peers in an emotive way by utilising "tunes". These tunes convey sentiments, like enjoyment, harmony, conscience and vibrations framed within creative arts (i.e. music, painting, writing). Emotive know-how transfer, peer-education and measures that are highly accepted by the target group, proved to be successful in changing attitudes and behaviour [3].

The general objective of the project is to improve the road safety level by reducing accidents, injuries and fatalities among young drivers through raising awareness of responsible and social behaviour in traffic.

The impact of the Safety Tunes methodology will be assessed with quantitative before and after surveys, measuring the attitudes of young drivers towards road safety issues before and after the implementation of the methodology. The project will also compare accidents, seriously injured and fatalities, as well as other road safety indicators within the age group of 15-24 in the Safety Tunes regions before and after the implementation. There will also be process evaluation with qualitative observations of the workshops including feedback of students and teachers. The aim is to optimize the training units and to see if the method has been accepted.

Another objective of the project is to transfer the method to other European countries. By evaluating the process of implementation in eight partner-countries, the extent of transferability will be demonstrated. To ensure transferability an input from the ITIM 5-D Model on intercultural differences will be used [4].

The specific objectives of the project are to:

- implement Safety Tunes in eight countries: Austria, Belgium, the Netherlands, Poland, Spain, Hungary, Czech Republic, Slovenia;
- reach 3,500 students through direct involvement in workshops;
- change attitudes towards risky behaviour of at least 70% of the students;
- reach 8,000 students indirectly, as audience of social-arts outputs (music, paintings, writing, etc.);
- reach 8,000 students through social media (Facebook / YouTube / Twitter / Soundcloud, etc.);
- decrease fatalities within Safety Tunes regions by 5% within the age group of 15-24;
- transfer the methodology to five other European countries [5].

2 Implementation of the Safety Tunes method

Safety Tunes is a methodology that contains a series of workshops reaching the target group of young drivers with a creative peer-approach. In each participating country, workshops are implemented in vocational schools by Safety Tunes teams to create road safety related messages through different types of artistic forms. Also, national and European competitions will be organized to motivate and award the best results [3].

Safety Tunes combines peer approach, an emotive know-how transfer and social art with road safety. Social art focuses on an active involvement of participants. The students have to deal with road safety topics in an interactive and creative way. If learned with creative methods, the content is more memorable. The methodology also demands the participants to reflect their own experiences and feelings, so that the emotional level is reached. Reaching the emotional level as well as learning from a peer, whose messages are more accepted than appeals made by institutions, have proven to be successful in changing attitudes and behaviour.

Safety Tunes teams guide student groups from different schools regarding the choice of theme and message, development of their social-art output and dissemination among peers and on social media. The team consists of road safety experts (consortium Partners), artists, teachers, social workers, students, etc. The workshops follow three steps [3]:

1. *Facts and Figures* – Students not only inquire about accident statistics, but are also confronted with the risks involved in cycling, using mopeds, motorcycles or cars. They learn about social behaviour in traffic, their responsibility towards other vulnerable road users and the risks of distraction. They also reflect their own experiences by e.g. using dilemma stories.
2. *Feel it!* – Production of social-art output. While working on messages, slogans, music and acting, the emotional level can be reached and a sustainable, responsible attitude can be anchored. Students also know best how they can convey their message in a direct and appealing way to their peers.
3. *Do it!* – Students are encouraged and motivated to present their art, talk about their experiences with friends and share the social-art output via social media (Facebook / Youtube / Twitter / Soundcloud, etc.).

It is expected that in each Partner country 20-60 workshops will be implemented in different regions and at least 24 social-art outputs with a road safety message will be produced. Planned social-art outputs include music videos, songs, paintings, sculptures, street performances, graffiti, theatre shows, short films, dilemma stories, poetry-slams, stand-up shows, etc.

The best and most effective social-art outputs (selected by social media and a jury) will be honoured in a national awards ceremony. The winner of the national competition will be invited to the European awards ceremony, that will take place during the final project conference in Warsaw in 2017 and during which the outcomes of different road safety actions in each country will be presented and discussed [5].

3 Evaluation of the project and its impact

One of the tasks within Safety Tunes Project is the evaluation of the Safety Tunes method and workshops, during which the ST method is implemented. The assessment consists of two parts:

- impact evaluation,
- process evaluation.

The impact of the Safety Tunes method is measured with quantitative surveys (with at least 3,000 questionnaires) measuring the attitudes of young people aged 15-24 attending vocational schools towards road safety issues before and after the implementation of the ST methodology. Additionally, there are questions about the general level of road safety knowledge [4].

The impact is also assessed by comparing accidents, fatalities and other performance indicators in ST Partners' countries before and after the implementation of ST.

The process is measured using specially developed questionnaires. This evaluation is carried out among students who participate in the ST workshops, trainers who run the workshops, school teachers who assist during the workshop (even if they do not actively participate) and ST project partners.

The process assessment with qualitative observations of Safety Tunes workshops aims to optimize the training units using ST and also to show whether the method has been accepted.

The evaluation methodology is mostly based on the pre and post assessment. Fig. 4 shows the whole process of assessment by process and impact evaluation.

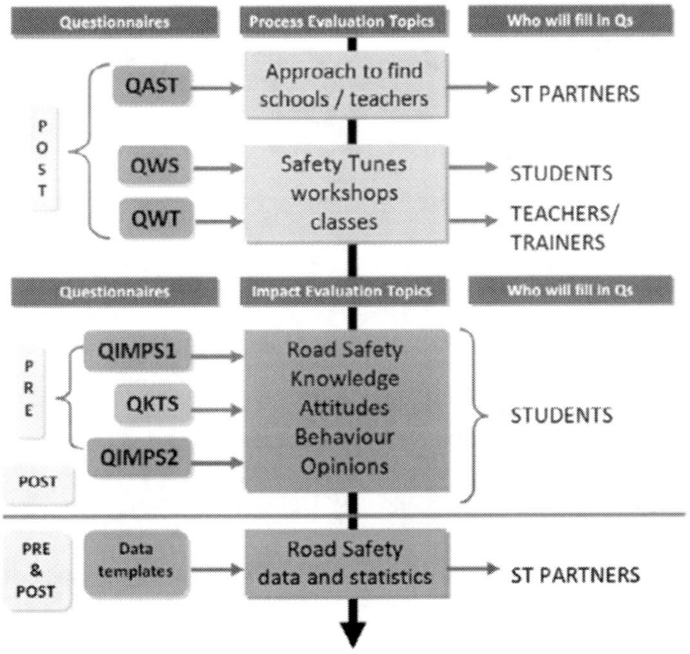

Fig. 1. Impact and process evaluation within Safety Tunes project [4].

3.1 Impact assessment

The overall objective of the impact assessment within the Safety Tunes project is to prove that the implemented ST method was effective in reaching the goal of change in attitudes/opinions of young people aged 15-24 attending vocational schools.

The impact evaluation in general:

- assesses the changes that can be attributed to a particular intervention (project, programme or policy),
- answers the question: how would outcomes such as participants' attitudes have changed if the intervention had not been undertaken,
- looks at the long-term, deeper changes that have resulted from the action.

The impact evaluation within the Safety Tunes project is targeted at vocational school students, who participate in the ST workshops during which the ST method is used. The outcomes of the pre assessment and post assessment will be then compared in order to show the effects of the project's impact on knowledge, opinions and perceived behaviour of the participating students.

The impact evaluation measures:

- change in attitudes/opinions of students regarding road safety issues,
- change in road safety data and performance indicators.

All data will be collected before and after the implementation of the ST method.

In order to assess a road safety situation, it is important to observe and monitor the number of road accidents and indicators such as road deaths by population. These data will be provided as soon as they become available. The source of the data will be the European database CARE – Community Road Accident Database and IRTAD – International Road Traffic and Accident Database, as well as EuroStat [4].

3.2 Process assessment

The main goal of process evaluation is to measure the actual development and implementation of a particular action and to establish whether the targets and implemented strategies have been achieved as planned.

Qualitative evaluation (by means of questionnaires) of the workshops after the implementation of ST method includes:
- feedback from students,
- feedback from trainers/teachers,
- feedback from ST Partners.

The aims are to:
- optimize the training units,
- show the acceptance of the ST method.

All the stakeholders (students, teachers/trainers and ST Partners) are asked to share their opinions/comments/remarks on the implemented ST workshop. This gives an overview on the work done with the target group and how the process is perceived by the target group, teachers and ST Partners.

Students and teachers fill in QWS and QWT post questionnaires and give their feedback on the ST workshop.

ST Partners are asked to share their experiences regarding the approach they used in order to find participating schools and teachers/trainers; who was contacted; etc. by filling in a QAST questionnaire [4].

All data received after the implementation of Safety Tunes workshops will be analysed and conclusions drawn in order to see how effective this method was in making vocational school students aware of different road safety problems among this group of users.

4 Conclusions

Methods such as Safety Tunes should become more popular and implemented on a wider scale in other regions as well. This innovative approach has been so far positively evaluated and the level of acceptance is quite high in the first year of implementation. The evaluation process is still in progress, as the project ends in 2017 and then the full results of the impact and process assessment will be available. All Partner countries run their activities and disseminate the project through various channels, such as the European Road Safety Charter and social media such as Facebook, Twitter, Soundcloud to gain more recognition and raise awareness of road safety problems among young road users.

References

1. European Road Safety Tunes, www.safetytunes.eu, (2016)
2. D. Jankowska-Karpa, J. Wacowska-Ślęzak, EU Project European Road Safety Tunes, BRD - Bezpieczeństwo Ruchu Drog. **2** pp. 15–18 (2016)
3. European Road Safety Tunes. Methodology Guidelines. Internal document, non public (Leuven, Begium, 2016)
4. European Road Safety Tunes. Impact Assesment Methodology. Internal document, non public (Warsaw, Poland, n.d.)
5. European Road Safety Tunes. Intermediary Report. Internal document, non public (Graz, Austria, 2016)

Comparison of requirements for location, maintenance and removal of road advertising between Polish and foreign regulations

Joanna Żukowska[1,*]

[1]Gdansk University of Technology, Faculty of Civil and Environmental Engineering, Gdansk, Poland

Abstract. The article gives an overview of Polish and international formal and legal requirements for roadside advertising and the relevant road safety impacts. The analysis focussed on outdoor advertising life cycle consisting of three stages: location, operation and removal of advertising. Experience of road authorities from Australia (Queensland), Republic of South Africa and the United Kingdom was collected. The article is part of a joint project "Development of Road Innovations" funded by the National Centre for Development and Innovation and the General Directorate For National Roads and Motorways.

1 Introduction

In times of free market economy we are inundated with information. Companies, organisations and stores use advertising to put their message across to as many future customers as possible. As well as using television, newspaper and radio ads, outdoor billboards are placed in generally accessible open space. With no legal regulations on where and how densely the ads can be located, the roadsides across the country, and in tourist destinations and major towns in particular, are brimming with all sorts of adverts. So much so that it interferes with the general signage such as street names, locations of government offices, landmarks or even road signs.

In legal terms, roadside advertising in Poland is already regulated. Using the roadside as a location for a billboard can only be done at the permission of the relevant road authority. What is not clear is the standards for placing advertising outside the road perimeter because the current legal regulations do not protect the interests of road users sufficiently to ensure road safety. The rulings of administrative courts suggest that road safety is important and should be considered when allowing roadside billboards [1].

Recently, public debate in Poland has voiced strong criticism towards the inundation of landscapes with advertising (along rural roads) and city aesthetics. Major cities have even set up departments to fight illegal and ugly advertising. As an example, Gdansk has its own City Aesthetics Department whilst Gdynia appointed the Municipal Arts Office to deal with advertising. Research and analyses on the relation between advertising visible from the road and road safety is scarce. With a dynamic increase in advertising presence, "visual noise" is created which is clearly bad for the aesthetics of the surroundings and causes chaos with understanding road and traffic signage. This may have an indirect effect on road risk. As we know from the literature visual noise is "constant background noise derived from a multitude

* Corresponding author: joanna@pg.gda.pl

of cues, interfering with or preventing the driver from processing the information from the cue significant to him" [2]. This means that a specific space is full of advertising making correct interpretation/understanding of road signs more difficult and distracting drivers. This phenomenon has been studied in other countries. A number of studies looked at how this problem affects driver behaviour. Countries that have significant achievements in studying the effects of advertising on driver behaviour include Australia, United States, Republic of South Africa and Canada. Extensive efforts were taken there to develop legal regulations and guidelines to help with a better control of outdoor advertising. There is a similar need in Poland.

2 Overview of international experience

Other countries have studied the phenomenon in extensive research on driver behaviour with Australia, United States, Republic of South Africa and Canada as the countries with the biggest achievements in studying the problem. They also started a major effort to create legal regulations and guidelines to help them control outdoor advertising in public space. This paper presents a synthesis of the analyses mostly from Australia (Queensland) [3], Republic of South Africa [4] and the United Kingdom [5]. The work relates to the so called life cycle of outdoor advertising (Fig. 1). It has three stages: advertising location, maintenance and removal. The paper describes the experience of road authorities regarding the application procedure, examples of application forms, areas where advertising is excluded and the permissible physical features of advertising.

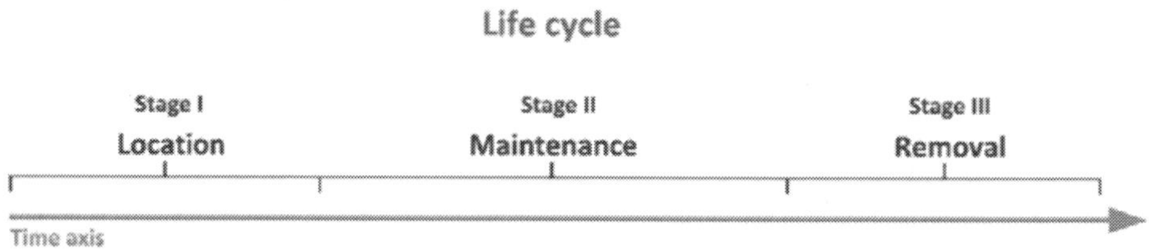

Fig. 1. Life cycle of advertising.

Requirements for advertising location. In Australia the state of Queensland requires applicants to receive a written permit from local authorities irrespective of whether the billboard is within the road perimeter or outside it. Local law sets out the requirements to be met by billboards. If they are to be situated within the road perimeter, the location must be authorised by the road authority. Its decision is binding. For a billboard to be placed within the road perimeter, the applicant must seek the permission of both the local authorities and road authorities. To give local authorities more powers in deciding the location on roads that are managed by national level road authorities, a guide was developed titled "Roadside advertising within and beyond the boundaries of state-controlled roads" which helps the local government to take on a greater role in managing roadside advertising on state-controlled roads [3].

Requirements for advertising maintenance. In the Republic of South Africa advertising that is permitted in a location should meet the following criteria for the purposes of maintenance [4]:
- be located at a height that discourages vandalism;
- be serviced on a regular basis – visual aspect;
- be maintained in good repair and in a safe condition.

Those responsible for maintaining the advertising in a safe and proper condition include both the owner of the advertising and the permitting body. To ensure that the criteria are met, it is recommended that the advertising be checked once a year. The land owner should have complete documentation of the location permit and, where the land owner does not own the advertising, he should receive a copy of the documents from the applicant. This is required when the Agency conducts checks later on. The Agency may annul the permit if the advertising does not meet or no longer meets the requirements set out in the permit.

Requirements for advertising removal. In the state of Queensland an advertising device may be removed if it poses a risk to road safety, hampers roadworks, if permit terms and conditions are violated or if the advertising is not licensed. The owner of the advertising is notified in writing about the order to remove the advertising. If the advertising is of low value (it is made of cardboard, paper, etc.), the relevant control body may remove the advertising immediately without notifying the owner. If the owner does not respond, the advertising is removed at the owner's cost by the Department or local authorities and penalties will apply. In the case of advertising placed on vehicles and trailers, if the vehicle is parked within the boundaries controlled by local government or the Department, and if the vehicle is there for displaying the advertising only, the vehicle should be removed immediately. In the case of advertising visible from the road, the procedure to remove it is similar.

3 Review of Polish experience

Poland's national laws do not address the location of advertising understood as expositional contents in too much detail. While there are regulations on structures in the Construction Law [6] and the Public Roads Acts [7] they only apply to structures that will be used to display advertising contents. If the advertising is not displayed on a structure within the meaning of Construction Law [6], this is not regulated. The historic buildings act is an exception [8]. There is nothing to regulate the shape, contents and ways to display advertising. Some mention can be found in the Landscape Act regarding the management of advertising display. This is a nation-wide regulation [9]. While the act helps to manage advertising, the risk is that advertising visible from the road will not be covered by local authorities from the perspective of road safety. Because the act allows municipalities to make their own decisions, the regulation may be enforced differently by the different municipalities which is not good for drivers. Road authorities work to standardise the parameters of a class or road category. In Poland a road authority may decide on what happens within the road perimeter but has hardly any say on what happens outside it.

Requirements for advertising location. Road safety is to be ensured in particular within the road perimeter. Roadside advertising must meet stringent regulations under the public roads act whose purpose is to create the conditions for an optimally safe road traffic and minimise the emergence of factors that could put safety at risk. Owners and road authorities are required to look after the road and the road perimeter to ensure that they can be used safely by road users. To fulfil this obligation, the authorities have powers to regulate road perimeter occupancy permits using their discretion. While discretion does not mean arbitrary decisions, authorities can choose between different solutions and weigh the goal of the law against public interest. This means that if the applicant meets the conditions of the road perimeter occupancy permit, the application is not necessarily approved automatically (c.f. verdict of the Supreme Administrative Court of 12 December 2008, II GSK 565/08). Road authorities are not allowed to permit an action, i.e. the siting of advertising, that is against the law. This conflict, however, cannot be based on subjective assessment. It must refer to the facts and the law. Road authorities have a duty to monitor the roads they manage and commission opinions and analyses and use materials developed by other organisations.

Requirements for advertising maintenance. Pursuant to Article 38.1 of the Public Roads Act [7] building structures and devices that have no relation to the road or traffic and are found within the road perimeter and do not pose risk to road traffic or hamper it and do not interfere with road authorities work, may remain unchanged. This regulation applies to building objects and devices that were found within the road perimeter on the effective date of the law or came to be there later because the road perimeter itself has been changed. The regulation includes a statutory permit of sorts with no time limits to occupy the road perimeter and no fees. It needs to be stressed that Article 38.1 of the Public Roads Act [7] applies to existing objects only in the road perimeter and it does not apply to objects that were placed there within the meaning of Article 40.1 and 2 of the Public Roads Act [7], such as advertising. The permitting body may take steps to eliminate the devices from the road perimeter by not renewing the permit to occupy the road perimeter. A previous road perimeter occupancy permit does not come with an automatic renewal, because the public road authorities have the right to decide about the road. Permits are for a limited time which is stated in the permitting decision. Once the validity expires, the road perimeter occupancy permit must be renewed to ensure the legality of the outdoor advertising. Just because an advertising device has been in a location for a long time and has had multiple consecutive permits does not mean that the permit will be renewed again.

Requirements for advertising removal. If the advertising device is placed within the road perimeter without the permit of the road authorities or if its temporary permit has expired and the applicant has not asked for a renewal or if the advertising is not built as designed, it will have to be removed and the owner of the advertising will pay a fine. The fine is a multiple of ten fees for a legal advertising device. In exceptional cases if the advertising complies with land use and building regulations, it may be legalised. A legalisation fee must be paid specified in Article 49.2 and Article 59f.1 of construction law [6]. It is often difficult to identify the owner of illegal advertising within a road perimeter. Orders to remove and disassemble advertising are issued by building supervision. Removal is done at the owner's cost. Areas outside the road perimeter are not regulated as regards the siting of advertising and its possible negative impact on road safety. Consequently, road authorities have no means to act. While they may report problematic advertising to building supervision, experience shows that responses to such instances differ from county to county. Some authorities will work to have the advertising removed, others will not see this is as road traffic hazard. With an abundance of roadside advertising in Poland, building supervision bodies cannot handle the majority of cases due to their capacity. In addition, the steps are taken under the code of administrative procedure which is lengthy and the penalties are low compared to the benefits gained from displaying the advertising. Some advertising owners are known to pay fines and continue to display the advertising or take it down when requested by the authorities only to put it up again because they know that a new administrative procedure will have to be launched and before any fines are paid, they will have benefited from the gains of the advertising.

4 Conclusions

Visual information in Poland is displayed irrespective of ownership boundaries or the boundaries of road perimeter. In the case of advertising facing the road (so that it is visible for road users) located outside the road perimeter, there is nothing road authorities can do to remove the advertising if it poses a road safety hazard. As a result, advertising visible from the road is not regulated as regards the location and operation and many organisations and institutions, both central and local have asked for specific rules to regulate this.

If the demands of local authorities, road authorities, building supervision, etc. are to be met, tools must be developed to help with planning, assessing, permitting and refusing to permit locations and forms of advertising visible from the road. Nation-wide rules and forms of how advertising contents visible from the road can be located and displayed must be made part of the legislation. This means having to:

- revise and complement the acts: road traffic law, public roads, building law and code of offences,
- develop and adopt an ordinance of the minister for transport on the technical conditions and siting rules for advertising visible from the road,
- enable local authorities to enact local laws to limit advertising in their area.

Before these documents are developed, driver behaviour must be studied as well as in-depth analyses must be conducted into the effects of driver distraction on road safety where advertising is displayed. An overview of international legal practice shows that the decisions to locate, control and remove advertising are mostly within the remit of local authorities. In the case of advertising visible from the road, road authorities are usually consulted. If advertising is to be placed within the road perimeter, the opinion of the road authorities is decisive. Advertising should meet the conditions set out in local law, acts, ordinances and guidelines on outdoor advertising. Administrative procedures are described in legal acts, while guidelines are prepared in the form of guides for advertisers and local authorities. Guidelines give a detailed description of the physical features of advertising and where it can be located. It is clear in international regulations that the shape and appearance of advertising devices cannot resemble that of official signs and road signs. The majority of cases ban the use of dynamic lighting and visual effects in advertising. Siting criteria are more rigorous where driver concentration must be ensured to take a decision such as before a pedestrian crossing, when joining traffic and approaching junctions. In Australia [10] special emphasis is put on advertising location with very detailed descriptions of methods for designating excluded areas. The Republic of South Africa [4] analyses advertising contents very carefully. Advertising should not have an effect on road safety and should be maintained as set out in the permit. In other countries unauthorised advertising is removed at the owner's cost. Permits are granted for a limited time and advertising may be removed if the permit is breached. The owner of the advertising device must ensure that the device is maintained in good technical condition and that it complies with the permit.

Studies of the literature have also shown that:
- many countries in Europe and worldwide regulate the location of advertising visible from the road and the forms of displaying advertising contents,
- while research on how advertising affects road safety is very complicated, all existing studies show that advertising attracts drivers' attention,
- studies have shown that variable and dynamic advertising attracts drivers' attention more often and for longer than conventional advertising,
- attracting attention and distracting is largely dependent on the contents and form.

Acknowledgements

Research conducted under project implemented within RID programme, funded by the Polish National Centre for Research and Development and the General Directorate for National Roads and Motorways in Poland, for the years 2016-2019.

References

1. T. Mackun, K. Jamroz, J. Żukowska, J. Rychlewska, Zagrożenia dla uczestników ruchu drogowego stwarzane przez reklamy widziane z drogi, Drogownictwo. **4–5** pp. 134–141 (2015)
2. J. Wachtel, T.V. Group, Wpływ na bezpieczeństwo nowej cyfrowej technologii wyświetlania w odniesieniu do zewnętrznych znaków reklamowych, **7** (2009)
3. Guideline: Roadside Advertising Guide, Edition 1.2 (Departament of Transport and

Main Roads, Australia, 2013)
4. The South African National Roads Agency limited repeal of certain regulations and making of regulations (The South African National Roads Agency, 2000)
5. Outdoor advertisements and signs: a guide for advertisers (Department for Communities and Local Government, 2007)
6. Act of 7 July 1994 on Construction Law, OJ 1994 No. 89 Pos. 414 (Poland, 2009)
7. Act of 21 March 1985 on Public Roads, OJ 2007 No. 19 Pos. 115 (Poland, 1985)
8. Act of 23 July 2003 on the protection of monuments and the guardianship of monuments, OJ 2003, No. 162, Pos. 1568 (Poland, 2003)
9. Act of 24 April 2015 amending certain acts on strengthening the tools for landscape protection, OJ 2015 Pos. 774 (Poland, 2015)
10. Advertising Signs Assessment Guidelines for Road Safety (Government of South Australia, 2014)

Bus bays inventory using a terrestrial laser scanning system

Katarzyna Bobkowska[1,*], *Jakub* Szulwic[1], and *Paweł* Tysiąc[1]

[1]Gdansk University of Technology, Faculty of Civil and Environmental Engineering, Gdansk, Poland

Abstract. This article presents the use of laser scanning technology for the assessment of bus bay geo-location. Ground laser scanning is an effective tool for collecting three-dimensional data. Moreover, the analysis of a point cloud dataset can be a source of a lot of information. The authors have outlined an innovative use of data collection and analysis using the TLS regarding information on the flatness of bus bays. The results were finalized in the form of colour three-dimensional maps of deviations and pavement type.

1 Introduction

Millions of people use roads every day all over the world. Roads, like many other structures, have an estimated durability of road types. In Poland a lot of the roads were built at the turn of the 20th and 21st c., especially for light cars [1]. Many of these roads carry traffic and heavy goods vehicles [2] which were not predicted when the road was first designed. It creates a lot of problems with technical conditions and the infrastructure must be improved. Treatments can be problematic, because restoring the original properties of the road requires workers to restrict traffic as cars may cause a lot of damage to the construction.

In small cities transport infrastructure is vital for regional growth. This is why it is important to maintain proper conditions to ensure optimal capacity. To support economic growth, new roads must be built. As indicated above, most of the roads were built at the turn of the 20th and 21st c. and a complete renovation is too expensive. To reduce the deterioration heavy vehicle traffic should be restricted by building ring roads. It is not possible to eliminate a degradation of roads because of public transport, where trams and buses operate in cities every day. Bus bays and the surfaces there are hardest hit by degradation. This paper shows a solution for making an inventory of bus bays and estimating their condition by using terrestrial laser scanning. This will support a proper analysis designed to find the best solutions to minimise the effects of heavy traffic on roads using the example of bus bays in Gdansk, Poland. While manual techniques could also be applied, they are often time consuming and the results are not accurate. To get a good understanding of the road conditions, it is better to use photogrammetry techniques.

2 Data acquisition and data processing

To acquire spatial data, the Riegl VZ-400 terrestrial scanning system was used (Fig. 1). A laser scanning system is an efficient tool for gathering spatial data with a very high accuracy

* Corresponding author: katbobko@pg.gda.pl

(5mm [3]). In addition to ranging, Light Detection and Ranging systems can provide information about the target and transmission path [4]. In that option we could indicate the lanes and measure the geometry of every pavement type.

Fig. 1. Riegl VZ-400.

The experiment was conducted by scanning positions at the beginning and end of the bus bay. To be remembered is the traffic. While noise from the cars is easy to remove, it can leave a shadow which could cause unexpected errors in analysing (for example errors when meshing the data) the scanned object. The results of the alignment of the scanned bays are presented in Fig. 2.

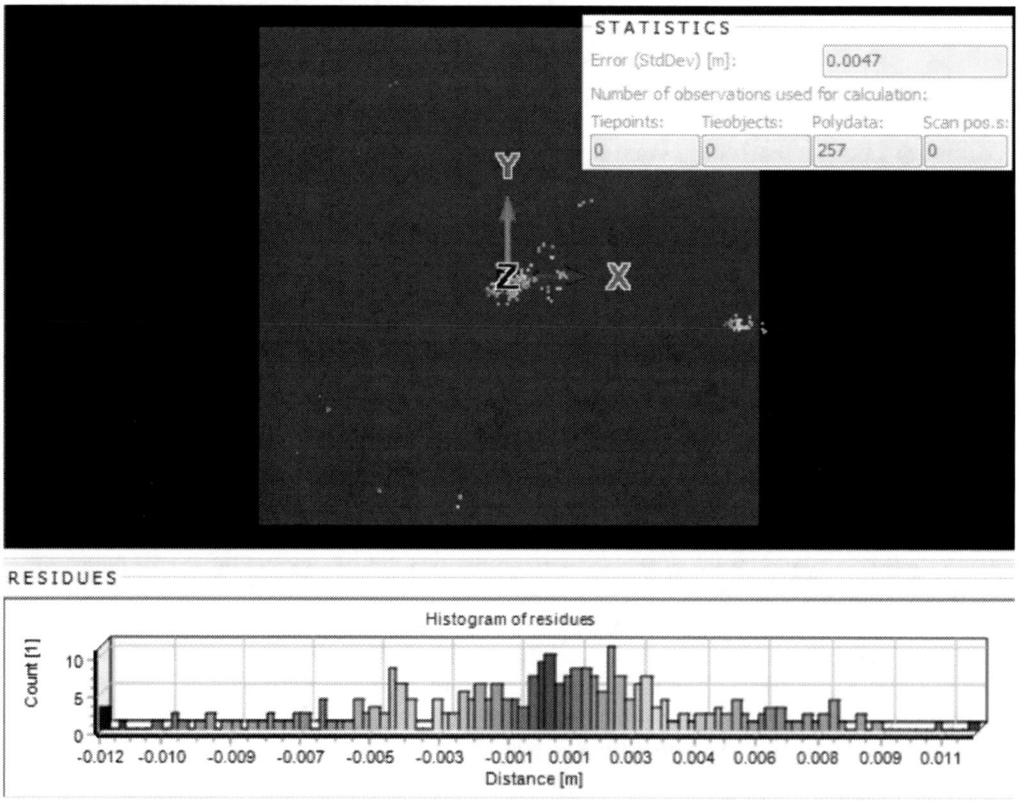

Fig. 2. Alignment of the scans results.

3 Classification of pavement condition

Flatness is one of the parameters which could be used to determine the level of service. It is a condition that affects driving comfort. The main requirement is a good, strong, flat, waterproof, durable and economic solution over the life of the object [5]. To meet the requirements, monitoring and evaluation must be conducted regularly to ensure that the proper method is used for improvements to the structure [6].

For the purposes of this paper, the authors chose the classification of road condition set out in attachment C to the system for evaluating the condition of pavements created by the General Director for National Roads and Motorways. The road condition in addition to its rut depth is shown in the table below.

Table 1. Road class in addition to depth of ruts.

Class	Pavement condition rate	Rut depth [mm]
A	good	<10
B	acceptable	11-20
C	not satisfactory	20-30
D	bad	>30

The laser scanning technology is a very effective tool as regards its ability to search for deformations [7] but it is undervalued in indicating ruts. As a result, the table presented above is not reliable. The reason for that is another measurement methodology described in the pavement condition evaluation system. The authors show an innovative solution which could be successfully used in addition to ruts depth in the future [8].

4 Experiment results

For the purposes of this paper, two different types of bus bays were scanned. The first one was built from asphalt and the second one from sett. The results of the two scans are presented below (Fig. 3, Fig. 4).

Fig. 3. Result of asphalt bus creek scanning.

Fig. 4. Result of sett bus creek scanning.

5 Bus creek condition

In order to assess the condition of the bus creeks, geometry analysis was conducted in addition to the individual segments of transport infrastructure objects. During the study, the CloudCompare software was used. The analytical process was as follows:

a) Separation of the cloud of points representing the portion of the bus creek, which will be subjected to analysis [9],
b) Fitting a plane on a point cloud (RMS fitting method),
c) Development of three-dimensional maps of deviations,
d) Visualization of maps in mesh form [10].

This approach allows for a quick geometry assessment of the road surface. Below are the results of the study:

a) the concrete blocks creek– Fig. 5,
b) the asphalt creek –Fig. 6.

Information on the selected points is displayed. In the yellow frame, the value of the deviation from the plane is indicated.

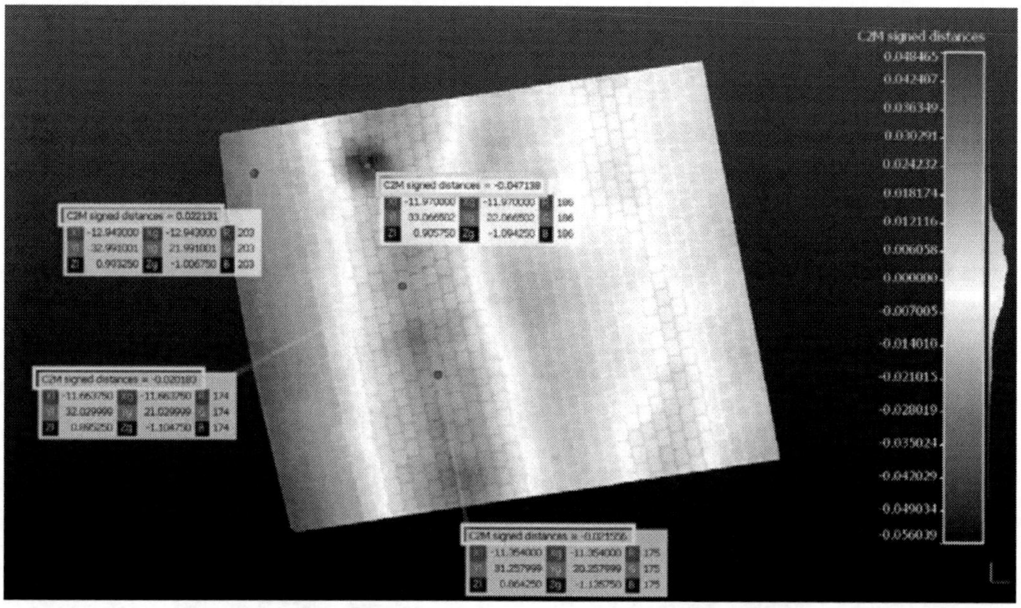

Fig. 5. Result of analysis of the segment (3.5 x 4 meters) creek with concrete blocks (units - meters).

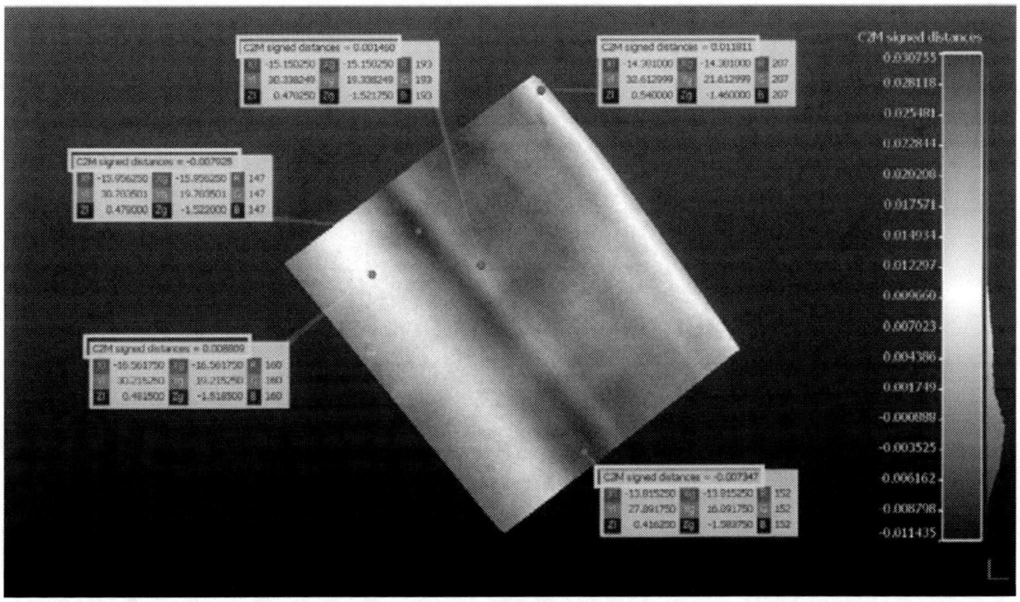

Fig. 6. Result of analysis of the segment (3.8 x 4.2 meters) asphalt creek (units - meters).

The drawings can be the basis for many conclusions about the state of the bus bays. The rating based on graphics is intuitive and makes it easy to identify ruts in an analysed area of road. The presented procedure can be widely used to analyse the geometry of flat objects when inventorying those objects.

6 Conclusions

In cities, the condition of roads is a very important factor because it determines driver and passenger comfort and safety [11]. The most degradable sections of the roads are bus bays. The authors analysed bus bays and the type of the pavement. To perform the experiment, the Riegl VZ-400 system was used, thanks to Apeks Company, Gdansk, Poland. Its innovative measuring method could be successfully used when searching for deformations. The experiment, described in this paper shows that of the two lay-bys taken into account, one is more deformed - a creek made of cobbles. In the case of this bay, the height differences are approximately 6 cm. In contrast, in the case of the bay made of asphalt, this difference is much lower - 2 cm. The proposed analytical process can be used in many other cases in transport construction (for example, to assess the condition of the surface geometry of roads, sidewalks or bicycle paths). Laser scanning technology has been used in transport construction for some time. An example may be a publication [12] describing the measurement of the clearance gauge of railways.

References

1. W. Kozłowski, Innovative Technological Solutions in Terms of Realizations Partial Renovation of Bituminous Roads Using SPRIDER, Procedia Eng. **161** pp. 503–510 (2016). doi:10.1016/j.proeng.2016.08.676
2. F. Li, S.S. Liao, M. Cai, A new probability statistical model for traffic noise prediction on free flow roads and control flow roads, Transp. Res. Part D Transp. Environ. **49** pp. 313–322 (2016). doi:10.1016/j.trd.2016.10.019

3. Riegl VMZ-400 Infosheet, www.riegl.com, (n.d.)
4. K. Bobkowska, A. Inglot, M. Mikusova, P. Tysi?c, Implementation of Spatial Information for Monitoring and Analysis of the Area Around the Port Using Laser Scanning Techniques, Polish Marit. Res. **24** pp. 10–15 (2017). doi:10.1515/pomr-2017-0015
5. J. Żukowska, Regional Observatory as a Part of Road Safety Information System, 11th Int. Conf. Transp. Syst. Telemat. (TST 2011). **239** pp. 409–416 (2011). doi:10.1007/978-3-642-24660-9_47
6. M. Budzyński, K. Jamroz, J. Pyrchla, W. Kustra, K. Pyrchla, Identifying the Effects of Selected Road and Roadside Parameters on Road Safety Using Geodetic Techniques, in: 2016 Balt. Geod. Congr. (BGC Geomatics), pp. 154–158 (IEEE, 2016). doi:10.1109/BGC.Geomatics.2016.35
7. P. Tysiac, A. Wojtowicz, J. Szulwic, Coastal Cliffs Monitoring and Prediction of Displacements Using Terrestial Laser Scanning, in: 2016 Balt. Geod. Congr. (BGC Geomatics), pp. 61–66 (IEEE, 2016). doi:10.1109/BGC.Geomatics.2016.20
8. B. Hejmanowska, W. Kaminski, M. Przyborski, K. Pyka, J. Pyrchla, Modern remote sensing and the challenges facing education systems in terms of its teaching, in: 7th Int. Conf. Educ. New Learn. Technol., pp. 6549–6558 (IATED, Barcelona, 2015)
9. A. Janowski, K. Nagrodzka-Godycka, J. Szulwic, P. Ziolkowski, Remote sensing and photogrammetry techniques in diagnostics of concrete structures, Comput. Concr. **18** pp. 405–420 (2016). doi:10.12989/cac.2016.18.3.405
10. D. Zai, Y. Guo, J. Li, H. Luo, Y. Lin, Y. Sun, P. Huang, C. Wang, 3D road surface extraction from mobile laser scanning point clouds, in: 2016 IEEE Int. Geosci. Remote Sens. Symp., pp. 1595–1598 (IEEE, 2016). doi:10.1109/IGARSS.2016.7729407
11. R. Bergel-Hayat, J. Żukowska, Road Safety Trends at National Level in Europe: A Review of Time-series Analysis Performed during the Period 2000-12, Transp. Rev. **35** pp. 650–671 (2015). doi:10.1080/01441647.2015.1030005
12. S. Mikrut, P. Kohut, K. Pyka, R. Tokarczyk, T. Barszcz, T. Uhl, Mobile Laser Scanning Systems for Measuring the Clearance Gauge of Railways: State of Play, Testing and Outlook, Sensors. **16** pp. 683 (2016). doi:10.3390/s16050683
13. A.I. Setianingsih, S. Sangaji, A. Setyawan, Road Maintenance and Rehabilitation Program Using Functional and Structural Assessment, IOP Conf. Ser. Mater. Sci. Eng. **176** pp. 12030 (2017). doi:10.1088/1757-899X/176/1/012030

MATEC Web of Conferences **122**, 04002 (2017) DOI: 10.1051/matecconf/201712204002
GAMBIT 2016

New Polish catalogue of typical flexible and semi-rigid pavements

Józef Judycki[1], *Piotr* Jaskuła[1,*], *Marek* Pszczoła[1], *Dawid* Ryś[1], *Mariusz* Jaczewski[1], *Jacek* Alenowicz[1], *Bohdan* Dołżycki[1], and *Marcin* Stienss [1]

[1]Gdansk University of Technology, Faculty of Civil and Environmental Engineering, Gdansk, Poland

Abstract. The paper covers the following topics important for the development of the new Polish Catalogue of typical flexible and semi-rigid pavements: reasons for preparing the new issue of the Catalogue of typical flexible and semi-rigid pavements, items introduced in the new issue, organise the terminology related to pavements, design traffic calculations and new equivalent axle load factors, new materials and technologies included in the Catalogue, classification of subgrades based on the soil material and drainage conditions, designing of lower layers and improved subgrade, designing of the main, upper layers.

1 Introduction

"The Catalogue of Typical Flexible and Semi-rigid Pavements" is one of the main design and construction guides used by road engineers in Poland. The previous issue of the Catalogue was published in 1997. Since then considerable technological developments in the area of road construction have been observed. They were accompanied by the introduction of new standards and technical requirements consistent with the requirements of the European standards. The requirements set out for road construction materials have also changed. New materials have been authorised for use in paving work. Traffic levels today are much higher than in the mid 1990s and gross weights and axle loads have also considerably increased. The dimensions and weights of vehicles are now limited by new regulations co-ordinated with the EU Directive. The changes include higher permitted loads transmitted on the pavement by single axles and by sets of axles. A weighing-in-motion scheme was implemented on the road network and the number of measurement sites is still increasing. This was accompanied by developments in pavement design methods and in testing of pavement structures and road construction materials. More attention is now paid to sustainable development, protection of the natural environment, limiting consumption of energy and recycling of materials. All these changes and developments called for preparing a new issue of the Catalogue. Hence, the team from the Department of Road Engineering of Gdansk University of Technology was employed by the Polish highway agency – GDDKiA to prepare a new edition of the Catalogue. After a few years of research the Catalogue was finalised in March 2013.

* Corresponding author: pjask@pg.gda.pl

2 What is new in the 2014 issue of the Catalogue

There are a number of differences between the new issue of the Catalogue, further referred to as the new issue, and the previous issue of 1997. Attention is drawn to the following main changes:

- Modified and more accurate definitions of terms relating to flexible and semi-rigid pavements.
- 30-year pavement design life adopted for motorways and trunk roads. For other roads the 20-year design life of pavement remained unchanged. A more accurate new calculation method of design traffic and introduction of a new traffic class - KR7 for design traffic much heavier than the current limits. These changes were caused by the growth of road traffic in Poland.
- Some minor changes in soil bearing capacity classification and the secondary deformation modulus E2 has been added for subgrade classification purposes. The new issue requires verification of the design assumptions by checking the bearing capacity of soil at the construction stage i.e. after stripping of topsoil - for cut sections or on completion of embankments - for fill sections.
- Three bearing capacity classes have been assumed on top of the lower pavement courses, under the base course. The class is specified in relation to calculated traffic class. The rules for using the separation and drainage layers have been organised.
- Various design and construction options are described for the lower pavement courses and improved subgrade.
- New materials and technologies have been introduced. They include: thin layer asphalt surfacing, open-graded porous asphalt, recycled and other man-made materials and materials bound with hydraulic binders. There are new requirements concerning materials, formulated in compliance with the European Standards.
- The typical pavement structures defined in the new Catalogue 2014 were designed with the application of the mechanistic-empirical method including new fatigue criteria. The results were compared to typical pavement structures used in other countries with similar weather conditions. The experiences gained over the years in Poland have also been taken into consideration.

3 New schematic diagram and terminology related to pavement structure

A big challenge during the work on the new issue of the Catalogue were serious discrepancies in the technical terms used in relation to the pavement courses in various Polish documents. This situation caused problems in the design, construction and invoicing phases of projects. The terminological studies resulted in some alterations in the terminology used so far. The final result is presented in Fig. 1. The pavement structure is divided into two layer packages - upper courses and lower courses. Improved subgrade is part of earthworks. Moreover, the base layer is divided into upper base layer and lower base layer.

<table>
<tr><td rowspan="6">Pavement structure</td><td rowspan="4">Upper pavement courses</td><td colspan="2">Wearing course</td></tr>
<tr><td colspan="2">Binder course</td></tr>
<tr><td rowspan="2">Base</td><td>Upper base layer</td></tr>
<tr><td>Lower base layer</td></tr>
<tr><td rowspan="2">Lower pavement courses</td><td colspan="2">Subbase</td></tr>
<tr><td colspan="2">Capping layer</td></tr>
<tr><td rowspan="2">Subgrade</td><td colspan="3">Improved (stabilised) subgrade</td></tr>
<tr><td colspan="3">Original soil in cut or embankment in fill of bearing capacity class G1-G4</td></tr>
</table>

Fig. 1. Schematic diagram and terms relating to courses of flexible and semi-rigid pavement structures and improved subgrade.

4 Traffic load calculation and design traffic classification

In the new issue of the Catalogue there are seven classes of traffic which might be applied in the design of a road pavement (from KR1 – lightest to KR7 – heaviest). The class is chosen on the basis of the design traffic. The design traffic volume is defined as the cumulative number of equivalent single axle loads (ESAL) of 100 KN per traffic lane during the whole design life. The design traffic is calculated with the following equation:

$$N_{100} = f_1 \cdot f_2 \cdot f_3 \cdot (N_C \cdot r_C + N_{C+P} \cdot r_{C+P} + N_A \cdot r_A) \quad (1)$$

where:
- N_{100} – design traffic being the cumulative number of equivalent standard axles of 100 kN per traffic lane during the design life,
- N_C, N_{C+P}, N_A – cumulative number of HGVs without trailers (C), HGVs with trailers or semitrailers (C+P) and coaches and buses (A) during the design life,
- r_C, r_{C+P}, r_A – load equivalency factors (LEF) to convert the numbers of HGVs without trailers (C), HGVs with trailers (C+P) and coaches (A) to the number of 100 kN ESAL,
- f_1 – load distribution factor of design lane, f_2 – lane-width factor, f_3 – longitudinal gradient factor.

In the new issue of the Catalogue the load equivalency factors r_C, r_{C+P}, r_A were determined on the basis of the available weigh-in-motion data. Weighing in motion (WIM) is a continuous process, providing complete data on heavy traffic, including vehicle type identification, axle loads, axle configuration, vehicle dimensions and speed. The pavement loading investigation was carried out using the measurement data obtained from weighing of over 4.2 million HGVs. The analysis of traffic included a determination of axle load and gross weight distributions, distribution by vehicle type, annual, weekly and daily traffic load distributions and the percentage of overloaded vehicles [1,2]. The most important element of the research was evaluation of the severity of vehicle influence on the pavement structure, represented by the load equivalency factors. Their values were calculated for each recorded vehicle using the following methods: AASHTO [3], fourth power law, French method [4] and method developed at the Gdańsk University of Technology [5]. Parameters determining the severity of the vehicle's action on pavement structure, appropriate to the method were taken into account, including: axle spacing (relevant to tandem and tridem axles), type of pavement (flexible or semi-rigid) and thickness of pavement courses and finally tyre-pavement contact stresses. The final values of load equivalency factors for vehicle classes were determined through static analysis of the load equivalency factors of individual vehicles, taking into account weight and axle load variations depending on the maximum legal load and class of road. Some safety margin was added to account for different load application parameters on different roads, possible vehicle overloading, future increase in gross weights and axle loads of vehicles as well as the dynamic effects.

Table 1. Comparison of the load equivalency factors provided in the 2014 and 1997 issues of the Catalogue of Flexible and Semi-rigid Pavements

Vehicle class	2014 issue				1997 issue
	Road type and the legal limit of single axle load				
	Motorways and trunk roads (115 kN)	National roads (115 kN)	Other roads (115 kN)	Other roads (100kN)	
HGV without trailer – **C type**	0.50	0.50	0.45	0.45	0.109
HGV with trailer – **C+P type**	1.95	1.80	1.70	1.60	1.245 1.950
Coaches and buses – **A type**	1.25	1.20	1.15	1.05	0.594

The new issue of the Catalogue introduces factors accounting for the influence of road geometry on design traffic, namely the lane width factor f2 and the longitudinal gradient factor f3. Their values depend on traffic lane width and longitudinal slope.

5 New road construction materials and technologies recently introduced in Poland

The main changes concerning construction materials and technologies in relation to the 1997 issue of the Catalogue [6] include:

- New asphalt mixtures authorised for wearing course construction: stone mastic asphalt (SMA) and porous asphalt (PA) applied in one or two layers [7].
- Wider use of partly crushed material for production of unbound base mixtures. So far, it was a common practice to specify only fully crushed material for road base construction.
- Use of cold-mix recycling for production of base mixtures (mineral-cement-emulsion mixture and mineral mixture treated with foamed bitumen) for pavements for KR1-KR4 traffic classes [8]. These materials were not specified as suitable for new constructions in the 1997 issue of the Catalogue [6].
- The required strength class for hydraulically bound materials for base and sub-base courses is defined by traffic class of the road concerned. Previously the strength requirement was the same for all traffic classes, but dependant on course function (base/subbase).
- Use of hydraulic road binders for treatment of aggregate and soil layers apart from the previously used binders, which were: cement, slag and fly ash. Hydraulic road binders combine the desirable characteristics of cement, fly-ash and lime and are currently among the most used soil and aggregate binding agents. By binding and reducing moisture of the stabilised material these materials give very good results in a relatively short time.
- Mixing-in-place as a recommended stabilisation treatment for hydraulically bound lower courses of pavement and improved subgrade. This was possible due to great popularity of modern mobile auger mixing machines, which ensure the required quality parameters of stabilised material.

The requirements relevant to fatigue resistance of pavement structure are given in the Catalogue [9] for most of the materials. In case of other requirements the reader is referred to relevant Polish regulations, which provide detailed specifications for all road construction materials. Individual pavement design is also allowed under certain circumstances, thus opening the door for a wider use of newly developed technologies.

6 Evaluation of the subgrade soil and ground water conditions

The new issue of the Catalogue [9] introduced a number of changes to the procedure of evaluation of the subgrade soil and ground water conditions. This has been caused mainly by the developments in subgrade testing and strengthening methods. The main changes include:

- Subgrade under earthen structure (i.e. embankments) is clearly distinguished from the subgrade under a pavement structure,
- New depths for analysing subgrade conditions under pavements: 2 m for checking ground water conditions and 1 m for checking soil conditions, measured from the underside of pavement structure.
- Secondary deformation modulus E2 was added to the previously used bearing capacity evaluation criteria (CBR, frost heave susceptibility and drainage conditions).
- Obligatory verification of the design input assumptions during the progress of construction by comparing the modulus E2 measured on the subgrade surface with the value adopted as input for design. A lower than assumed value of E2 obtained in such verification requires re-designing of the lower pavement courses.

7 Designing the lower courses of pavement and the layer of improved subgrade

The Catalogue issue of 1997 gave a very limited choice in this question, which was generally between treatment of subgrade with cement or replacement of soil. It was established in a review of methods used in other countries that engineers are allowed to use different stabilisation systems which are appropriate for traffic loads and the required value of modulus on top of the subgrade. This way a stabilisation technique can be chosen which is the most appropriate for the local soil and water conditions. An example of a deficiency in the previous issue of the Catalogue (1997) was prescribing the same stabilisation depth and strength for two different values of secondary deformation modulus after treatment (100 MPa and 120 MPa) and treating G1 bearing capacity class as equivalent to meeting these requirements. Accordingly, no treatment was needed for subgrades built of materials such as fine sand. In the new Catalogue these provisions were verified and the above-mentioned simplifications and deficiencies were eliminated. The minimum value of E2 measured on top of the improved subgrade layer was introduced as $E2 \geq 50$ MPa. The value of E2 required on top of the lower pavement courses: sub-base and/or capping layer depends on the traffic class of the road. These are: $E2 \geq 80$ MPa for traffic classes KR1-2, $E2 \geq 100$ MPa for traffic classes KR3-4, and $E2 \geq 120$ MPa for traffic classes KR5-7.

The BISAR computer programme was used to evaluate and design different lower pavement course systems, represented by an elastic layered half-space model. For each layer appropriate values of E-modulus and Poisson ratio were specified. Elastic deflections on the top of the improved subgrade and on the top of the lower courses of pavement were determined for each analysed system. The value of equivalent modulus of elasticity Eeq on the top of the analysed layered systems were calculated using the Boussinesq's equation. Computations were conducted for 14 different systems of lower courses of pavement and improved subgrade and for all bearing capacity classes of subgrade - from G4 to G1. Some typical arrangements (types 1-4) for traffic categories KR5-7 and G4 bearing capacity class are presented in Fig. 2.

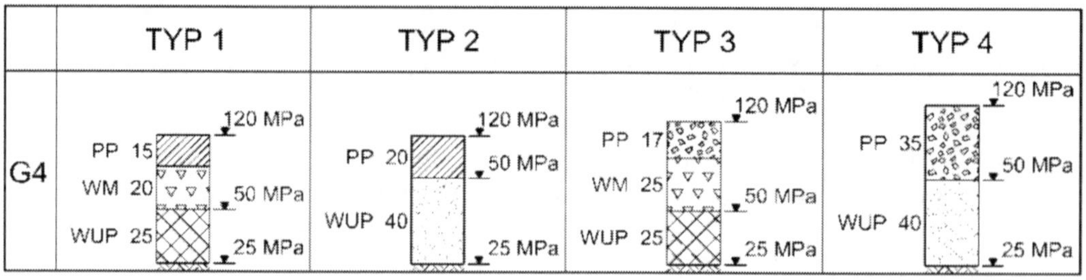

Fig. 2. Some typical arrangements (types 1-4) for traffic categories KR5-7 and G4 bearing capacity class of subgrade (PP – sub-base, WM – capping layer, WUP – improved subgrade layer)

The new issue of the Catalogue specifies thicker solutions as compared to the 1997 issue which, however, are in compliance with those currently used in other countries.

8 Upper courses of pavement

Taking into account the developments in the use of paving materials since the date of publication of the design methods recommended in the previous Catalogue it was decided that the use of Asphalt Institute [10], Shell [10] and AASHTO 1993 [3] methods will be limited to preliminary calculations and the results will be treated as supplementary. For a further application the fatigue criteria from the mechanistic-empirical methods AASHTO 2004 [11][12] and (partly) from the French method [13][14] were selected. Due to a lack of detailed data from laboratory testing of materials it was not possible to use the subgrade strain criterion from the AASHTO 2004 and the previously used subgrade deformation criterion of the Asphalt Institute method was chosen instead.

The stress and strain calculations in the pavement were carried out according to the theory of elastic layered half-space. An assumption that a single axis transmits the load through two single wheels, represented by circular contact area of P=50 kN load and q=850 kPa tyre-pavement contact pressure was assumed in the calculations as appropriate for contemporary heavy goods vehicles.

The design criteria used for flexible pavement design were: bottom-up fatigue cracking of asphalt layers, according to the latest AASHTO M-ENPDM method of 2004 (see equations 2 and 3) [11] and permanent deformation according to the Asphalt Institute method of 1982 [10] The number of load applications until the development of fatigue cracking in asphalt layers was calculated with the following equation:

$$N_f = 7{,}3557 \cdot (10^{-6}) \cdot C \cdot k_1' \cdot \left(\frac{1}{\varepsilon_t}\right)^{3{,}9492} \cdot \left(\frac{1}{E}\right)^{1{,}281} \tag{2}$$

where:

- N_f – number of load repetitions until the development of fatigue cracking on 50% of the overall surface area of traffic lane,
- k_1' – calibration constant depending on the asphalt layer thickness and type of fatigue cracking,
- ε_t – tensile strain at the critical point in the vertical cross-section of pavement, -,
- E – stiffness modulus of asphalt layer, MPa,
- C – coefficient depending on the volumetric parameters of asphalt mixture calculated as $C = 10M$, in which M is calculated as follows:

$$M = 4{,}84 \cdot \left(\frac{V_b}{V_a+V_b} - 0{,}69\right) \tag{3}$$

where:
- V_b – effective content of bitumen, % v/v,
- V_a – air voids content, % v/v.

The AASHTO 2004 criterion represented by equations (2) and (3) is based on the number of load repetitions until the development of fatigue cracking on 50% of the overall surface area of the traffic lane. However, by rearranging the above equations it is possible to calculate the number of load applications for any level of development of fatigue cracks. The following cracking severity levels were adopted for the purpose of the performed analyses: 10-20% for pavements with unbound base and 5-10% for full-depth asphalt pavements.

According to the conduced calculations and analyses all thicknesses of the upper pavement courses are sufficient to fully cover the load ranges in the respective traffic classes of roads. In comparison to the Catalogue issue of 1997 the new issue of 2014 provides more typical options for the design of upper courses in pavement structures with unbound base. What is completely new are typical pavement structures with cold recycled base and the use of porous asphalt in bituminous courses. The thicknesses of the upper courses of typical pavement structures are presented in Table 2.

Table 2. Thicknesses of the upper courses of typical pavement structures given in [9]

No.	Type of pavement	Thickness of the upper courses of pavement, cm							
		Pavement course	Traffic class, KR						
			1	2	3	4	5	6	7
1.	Flexible	Asphalt layers	9	12	16	20	24	28	30
		Unbound base, $C_{90/3}$	20	20	20	20	20	20	20
2.	Flexible	Asphalt layers	9	12	16	20	24	28	30
		Unbound base, $C_{50/30}$	22	22	22	22	22	22	22
3.	Flexible	Asphalt layers	9	12	-	-	-	-	-
		Unbound base, C_{NR}	25	25	-	-	-	-	-
4.	Flexible	Asphalt layers and base	14	18	22	26	30	34	36
5.	Flexible	Asphalt layers	8	12	12	16	-	-	-
		Mineral/cement/ emulsion/ FBit base	15	15	20	20	-	-	-
6.	Semi-rigid	Asphalt layers	9	11	15	18	20	22	24
		Hydraulically-bound base	18	20	20	22	22	24	24
7.	Semi-rigid	Asphalt layers	9	11	-	-	-	-	-
		Hydraulically treated soil base	18	20	-	-	-	-	-

9 Conclusions

The new Catalogue of typical flexible and semi-rigid pavements [9] provides new load equivalency factors (to convert vehicles to equivalent single axle load applications), defines a new traffic category, extends the design life period for motorway and trunk road pavements, considerably modifies the existing soil subgrade treatment methods and adds new ones, introduces new materials, including recycled materials for use in the construction of upper and lower courses of pavement and for improved subgrade. Moreover, it includes provisions intended to reduce considerably the problem of reflective cracking in semi-rigid pavements.

Safety margins are applied to the following issues included in the Catalogue: traffic calculation method, mechanical properties of asphalt and hydraulically-bound base layers, methodology of adopting a standard Catalogue pavement structure included in the Catalogue and selection of construction tolerances.

The new issue of the Catalogue offers a much wider choice of different systems and construction methods allowing for their adjustment to suit local conditions.

References

1. D. Ryś, J. Judycki, P. Jaskuła, Determination of Vehicles Load Equivalency Factors for Polish Catalogue of Typical Flexible and Semi-rigid Pavement Structures, Transp. Res. Procedia. **14** pp. 2382–2391 (2016). doi:10.1016/j.trpro.2016.05.272
2. D. Ryś, J. Judycki, P. Jaskuła, Analysis of effect of overloaded vehicles on fatigue life of flexible pavements based on weigh in motion (WIM) data, Int. J. Pavement Eng. **17** pp. 716–726 (2016). doi:10.1080/10298436.2015.1019493
3. AASHTO, Guide for Design of Pavement Structures (AASHTO, USA, 1993)
4. J.F. Corte, et al., Conception et dimensionnement des structures de chaussee (LCPC, 1994)
5. J. Judycki, A. Grajewska, et al., Recommendations for calculating the load equivalency factors for flexible and semi-rigid pavements (Gdańsk University of Technology, Highway Agency (GDDKiA), Gdańsk, 2006)
6. Catalogue of typical flexible and semi-rigid pavements (IBDiM, GDDKiA, Warsaw, 1997)
7. B. Świeczko-Żurek, P. Jaskuła, J.A. Ejsmont, A. Kędzierska, P. Czajkowski, Rolling resistance and tyre/road noise on rubberised asphalt pavement in Poland, Road Mater. Pavement Des. **18** pp. 151–167 (2017). doi:10.1080/14680629.2016.1159245
8. B. Dołżycki, M. Jaczewski, C. Szydłowski, The Influence of Binding Agents on Stiffness of Mineral-cement-emulsion Mixtures, Procedia Eng. **172** pp. 239–246 (2017). doi:10.1016/j.proeng.2017.02.103
9. J. Judycki, P. Jaskuła, M. Pszczoła, J. Alenowicz, B. Dołżycki, M. Jaczewski, D. Ryś, M. Stienss, Catalogue of typical flexible and semi-rigid pavements (Gdańsk University of Technology, 2013)
10. J. Judycki, Comparison of fatigue criteria used in the design of flexible and semi-rigid pavements in relation to the new issue of the Polish Catalogue of flexible and semi-rigid pavements (Part 1), Road Eng. **1** pp. 5–18 (1999)
11. J. Judycki, Building and calibration of models to investigate fatigue cracking in bituminous road pavements designed with the mechanistic-empirical method AASHTO 2004, Drog. I Most. **4** pp. 31–53 (2011)
12. J. Judycki, Models to investigate fatigue cracking in bituminous road pavements designed with the mechanistic-empirical method AASHTO 2004, Road Eng. **11** pp. 343–347 (2011)
13. M. Jaczewski, J. Judycki, Adaptation of the French fatigue criteria for designing of flexible pavements in Poland, Road Eng. **5** pp. 164–170 (2012)
14. M. Jaczewski, J. Judycki, Application in Poland of the French method for designing bituminous pavements including hydraulically bound roadbase, Road Eng. **6** pp. 199–203 (2012)

MATEC Web of Conferences **122**, 04003 (2017) DOI: 10.1051/matecconf/201712204003
GAMBIT 2016

Experimental road signs in a simulated environment research programme – experiment procedures and assumptions

Mikołaj Kruszewski[1,*], *Monika* Ucińska[1], *Ewa* Smoczyńska[1], *Paweł* Budziszewski[2], and *Andrzej* Grabowski[2]

[1]Motor Transport Institute, Warsaw, Poland
[2]Central Institute for Labour Protection – National Research Institute, Warsaw, Poland

Abstract. The application of experimental signs on public roads may cause some risks due to driver misunderstanding or misinterpretation, especially when seen for the first time. To minimize this risk, driver reaction and sign understanding can be tested in a safe and relatively cheap simulation-based environment. The consortium running the project "Experimental road marking and its effect on road user behaviour" suggested a similar methodology. The project included a simulation-based research program. The consortium proposed a set of experimental signs which have a strong potential for improving road safety and traffic conditions. The paper outlines the simulation-based research programme involving experimental signs and a proposed experiment procedure. Simulation scenarios and procedures were selected to achieve the required message and goals of implementation. In addition, an analysis is proposed of the factors and indicators of each sign and its effects.

1 Introduction

Experimental signs are generally used to draw the driver's attention to a new problem or to restrict a certain behaviour. Because experimental signs are not commonly used, drivers may be faced with a sign for the first time in their life. Therefore, experimental signs on public roads may cause some risks due to driver misunderstanding or misinterpretation, affect traffic flow and lead to dangerous situations on the road. To minimize this risk and understand its influence, a simulated driving environment is used before the new signs are deployed in a real road environment. The consortium running the "Experimental road marking and its effect on road user behaviour" project, suggested a similar methodology. In the project, simulation-based experiments for the assessment of experimental signs will be introduced.

This approach was inspired by the promising results of other research teams that used simulation environments to assess how road signs influence the driver. Research conducted by Charlton [1], Rosey and Auberlet [2] or Ding et al. [3] shows that a driving simulation environment is suitable for conducting such research. Using the simulator we can evaluate driver behaviour in a safe environment and see whether the objectives underlying the establishment of a sign will be achieved. The research and the authors' earlier work were the

* Corresponding author: mikolaj.kruszewski@its.waw.pl

basis for proposing such a methodology. The procedures and assumptions of the proposed research are described in the chapters that follow.

1.1 Experimental signs chosen for the simulation-based survey

During the first phase of the project members of the Project consortium selected several experimental signs that should be the subject of the survey during the project. These signs included both horizontal signing and markings and vertical signs. Based on literature review and research results of other international research teams, the selected signs were tested in a simulation-based environment.

The following sign types were selected:

- horizontal markings which are the optical means of speed reduction at entries to intersections, junctions and ramp lanes
- horizontal curve markings with additional warning signs
- horizontal markings for speed reduction
- markings indicating the need to keep a safe distance
- extension of non-standard horizontal curve markings.

Some of the above signs are tested based on good practices and implementation models from other counties. The simulation-based experiment is in this case conducted to verify the influence of the sign on Polish drivers. Other signs, however, have different models of implementation in different countries. Especially in the case of vertical signs, a similar meaning of the sign could be shown using a different pictogram design in different countries. In such cases, the consortium agreed to perform survey-based research, to choose the most understandable sign pictogram.

1.2 Survey-based research to choose sign pictograms

Selecting the most understandable sign pictogram is one of the key elements when preparing for a simulation experiment. The drivers who will take part in the study will see these signs for the very first time. The meaning of the sign should be clear, and its comprehension should not influence their driving performance.

Based on literature review, the open-ended survey method based on ISO 9186-1 [4] was proposed. The ISO 9186-1 standard is dedicated to testing non-verbal presentation of information in public areas and specifies methods for assessing the comprehensibility of graphical symbols [4]. The method is used for evaluating the share of people who can correctly understand the meaning of a symbol or sign.

At least three designs of signs for each of the meanings were used in the survey. The designs were prepared according to ISO 7010 [5], dedicated to the preparation of safety signs in public areas, and in accordance with national law on graphic symbols to be used in road signs in Poland [6].

For each sign two questions are provided:

- What action should you take in response to this symbol?
- What do you think is the meaning of this symbol?

The respondent should answer both questions with a few words describing what they thought first when they saw the symbol. Figure 1 shows an example from the questionnaire.

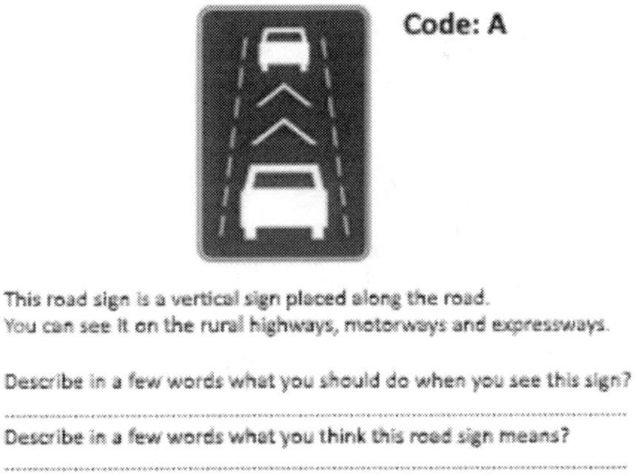

Fig. 1. Example of the survey showing a proposed sign design – horizontal road markings indicating the need to keep a safe distance to the vehicle in front.

Every response was categorized into one of five possible categories depending on the degree of correctness. The survey is planned for at least 300 respondents. The results should suggest which of the proposed sign designs is easiest to understand. The signs were used in simulation-based experiment.

2 Experiment procedures

The simulation-based experiment was conducted in accordance with the internal rules and good practices of the Driving Simulation Laboratory of the Motor Transport Institute. The simulation-based experiments need to comply with the safety procedures that include e.g. simulator sickness prevention rules or adaptation period. The experiment procedure will be in line with these rules providing suitable breaks for the participants and a corresponding sequence of tests. The whole experiment will be divided into relatively short simulation rides depending on the final number of road situations to be tested. Figure 2 shows the universal procedure scheme used in Simulation Laboratory.

Fig. 2. Example of a Simulation Laboratory research procedure used for experiments divided into two test rides.

The procedure applied in Simulation Laboratory helps to control the occurrence of simulation sickness and takes into account the necessity for the driver to adapt their skills to the specifics of driving a car in a simulation environment. A similar procedure was implemented successfully in a previous research e.g. EYEVID [7] and GEMS [8,9].

2.1 Apparatus

For the simulation-based experiment the project consortium plans to use two research tools. First, the AS1200-6 simulator. It is a high fidelity simulator built using the Opel Astra 4 cabin situated on a motion platform with 6 degrees of freedom of movement. The simulation environment is The SimWorld provided by AutoSim. The simulator records all data regarding the use of steering equipment.

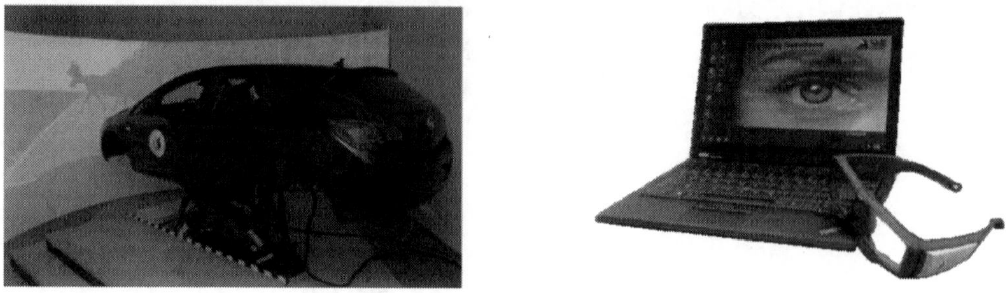

Fig. 3. Research tools intended to be used in the experiments. Left: AS1200-6 simulator. Right: SMI Eyetracking Glasses 1.8 mobile eyetracker.

Due to the project scope, the consortium decided to use the eyetracker to register the visual activity of the participants. The mobile eyetracker EyeTracking Glasses 1.8 of the SensoMotoric Instruments will be used. This tool should provide information about the time spent on sign observation, an indicator of a possible driver distraction.

2.2 Survey assumptions

Literature review provided information about the most popular parameters analysed in research regarding the influence of road infrastructure on the driver. Most effects are observed in lateral or longitudinal steering behaviour depending on the type of sign used in research. The following parameters were indicated to be recorded during experiments:
- Lane excursion
- Lateral behaviour
- Headway
- Time to collision
- Gas pedal steering
- Steering wheel reversal rate
- Speed.

All the above parameters would be registered and analysed during simulation-based research. For each parameter the analysis of the behaviour will be conducted on the section before the sign, in the area of the sign's impact and on the section after the sign.

For each of the signs a suitable road situation will be presented. The driver will drive in two similar road situations, one with the experimental sign, and one without it or with the standard road sign. Figure 4 shows an example of an experimental road situation.

Fig. 4. Example of an experimental road situation. Left: road curve with classical signing. Right: road curve with an experimental sign supplementing the classical sign.

The effect of the experimental sign will be compared to the standard sign in the same road situation. The comparison should suggest whether the expected results of the experimental sign have been achieved and whether the sign does not induce inadequate or dangerous behaviour. At least 50 drivers (different age, sex, and experience) are planned to participate in the experiment.

2.3 Control of simulator sickness occurrence

Simulation sickness control is a common problem for all research conducted in simulation environments. As indicated by Cobb et al. [10] simulator sickness is a problem that may:

- affect research results,
- cause that some participants may not be able to complete the tests carried out in the driving simulator.

To evaluate the occurrence of symptoms of simulator sickness the survey tool called RSSQ (Revised Simulation Sickness Questionnaire) was introduced. The RSSQ used in Simulation Laboratory of the Motor Transport Institute is a direct implementation of RSSQ developed by Kennedy et al. [11]. The survey includes 29 questions about the occurrence of the symptoms. Responses are evaluated in three categories of nausea, occulomotoric and disorientation symptoms.

The RSSQ is a post hoc method of simulation sickness evaluation. It helps to check how research results are influenced by helping to eliminate from the analysis the results of participants who may suffer significant inconvenience because of the symptoms. The results of these drivers are removed from the analysis.

SSQ1 Respondent code ....

Respondent's comfort (before driving simulator – adaptation)

The questionnaire aims to check what is your current comfort. Mark with a cross (X) which of the symptoms given below refers to your feeling status. Only one answer per line is allowed.

LP	Symptoms	None	Slight	Moderate	Severe
1	General discomfort				
2	Fatigue				
3	Tiredness				
4	Somnolence				
5	Headache				
6	Eyestrain				
7	Difficulty focusing				
8a	Increased salivation				
8b	Dry mouth				
9	Sweating				
10	Nausea				
11	Difficulty concentrating				
12	Depression				
13	Disorientation				
14	Blurred vision				
15a	Daze at eyes opened				
15b	Daze at eyes closed				
16	Dizziness				
17	Flashes of memory				
18	General weakness				
19	Need to take a breath				
20	Stomach discomfort				
21	Loss of appetite				
22	Increased appetite				
23	Need to defecate				
24	Sense of confusion				
25	Feeling of bouncing from stomach				
26	Vomiting				
27	Others:				

Fig. 5. Revised Simulation Sickness Questionnaire.

3 Conclusions

The process of implementing new road signs must take road safety as a priority. The influence of experimental signs on drivers depends on different contexts (e.g. cultural, organizational, legal, etc.) and needs to be carefully examined before the proposed signs are introduced on the roads. Surveys and a good class driving simulator seem to be the best solution for this purpose due to the controlled and repeatable simulation environment as well as precise driving data registered during the drive.

In the "Experimental road marking and its effect on road user behaviour" project a detailed research procedure was designed in order to deal with the presented topic. The method is based on the literature, international standards and good practices of members of the Project Consortium and provides a two-step procedure for choosing the best-fitting road signs. A simulation-based driving experiment, preceded with selecting the best signs evaluated in an online survey, will provide a sufficient amount of data to evaluate the change in driver's behaviour when faced with the new signs. This procedure is easy to implement and cost-efficient and could significantly simplify the process of road sign implementation, taking into account all accompanying factors.

Acknowledgements

The project "Experimental road marking and its effect on road user behaviour" is financed by the National Centre for Research and Development (NCBiR) and the General Directorate for National Roads and Motorways (GDDKiA), grant agreement DZP/RID-I-36/5/NCBR/2016.

References

1. S.G. Charlton, The role of attention in horizontal curves: A comparison of advance warning, delineation, and road marking treatments, Accid. Anal. Prev. **39** pp. 873–885 (2007)
2. F. Rosey, J.-M. Auberlet, Trajectory variability: Road geometry difficulty, Saf. Sci. **50** pp. 1818–1828 (2012)
3. H. Ding, X. Zhao, J. Rong, J. Ma, Experimental research on the effectiveness of speed reduction markings based on driving simulation: A case study, Accid. Anal. Prev. **60** pp. 211– 218 (2013)
4. Graphical symbols – Test methods – Part 1: Method for testing comprehensibility. ISO Norm 9186-1 (2014)
5. Graphical symbols -- Safety colours and safety signs – Safety signs used in workplaces and public areas. ISO Norm 7010 (2011)
6. Regulation of the Ministers of Infrastructure and Internal Affairs and Administration of 31 July 2002 on road signs and signals. OJ 2002 no. 170 pos. 1393 (Poland, 2002)
7. M. Kruszewski, P. Razin, M. Niezgoda, E. Smoczyńska, T. Kamiński, Analiza efektów oddziaływania symulatora na powstawanie choroby symulatorowej w badaniach kierowców, Syst. Logistyczne Wojsk (in Publ. **44** (2016)
8. M. Niezgoda, T. Kamiński, M. Kruszewski, A. Tarnowski, Self-reported drivers' behavior: an application of DBQ in Poland, J. Kones Powertrain Transp. (2013)
9. M. Niezgoda, A. Tarnowski, M. Kruszewski, T. Kamiński, Towards testing auditory–vocal interfaces and detecting distraction while driving: A comparison of eye-movement measures in the assessment of cognitive workload, Transp. Res. Part F. **32** pp. 23–34 (2015)

10. S. Cobb, S. Nichols, A. Ramsey, J. Wilson, Virtual reality-induced symptoms and effects (VRISE), Presence. **8(2)** pp. 169–186 (1999)
11. R.S. Kennedy, N.E. Lane, K.S. Berbaum, M.G. Lilienthal, A simulator sickness questionnaire (SSQ): A new method for quantifying simulator sickness, Int. J. Aviat. Psychol. **3(3)** pp. 203–220 (1993)

Searching for road deformations using mobile laser scanning

Jakub Szulwic[1,*], and *Paweł* Tysiąc[1]

[1]Gdansk University of Technology, Faculty of Civil and Environmental Engineering, Poland

Abstract. Millions of people use roads every day all over the world. Roads, like many other structures, have an estimated durability. In Poland a lot of the roads were built at the turn of the 20th and 21st c., especially for light cars. Many of these roads carry traffic and heavy goods vehicles which were not predicted when the traffic was first estimated. It creates a lot of problems with technical conditions and the infrastructure must be improved. Treatments can be problematic, because restoring the original properties of the road requires workers to restrict traffic as cars may cause a lot of damage to the construction. In the article the authors present a method for estimating the condition of a road using the MLS (Mobile Laser Scanning) measurement technique. It is based on a mobile platform and equipped with the Riegl VMZ-400 scanning system. Post-processing of the data constrain to extract the scan lines and road's condition analysis in addition to estimate its parameters. In conclusion, the authors present the advantages and disadvantages of Mobile Laser Scanning in addition to the paths. Moreover, tell the possibility of the factors determine which describe the security level of the roads.

1 Introduction

Millions of people use roads every day all over the world. Roads, like many other structures, have an estimated durability. In Poland a lot of the roads were built at the turn of the 20th and 21st c., especially for light cars [1]. Many of these roads carry traffic and heavy goods vehicles which were not predicted when the traffic was first estimated. It creates a lot of problems with technical conditions and the infrastructure must be improved. Treatments can be problematic, because restoring the original properties of the road requires workers to restrict traffic as cars may cause a lot of damage to the construction.

Another issue other than repairing roads, are traffic accidents. Most of them are caused by incompetence of the drivers who exceed speed limits. And even when the speed limit is respected by the drivers, the technical condition of the roads combined with bad atmospheric conditions could cause very problematic situations. To reduce accidents, it is very important to implement the safety policy proposed by the transportation office. To create a safety plan a lot of factors have to be taken into consideration such as the environment. Noise pollution in cities is a problem which was created by roads. The solution is to predict the level of traffic noise. It is possible to study the average level of traffic over a period of time (for example traffic volume or average traffic speed) [2]. Because the technical condition of roads is so important, we must remember that roads behave dynamically. When developing a proper

* Corresponding author: szulwic@pg.gda.pl

safety plan and action plans, the condition of the roads cannot be ignored. Apart from environmental factors, an inventory has to be conducted. It is possible to measure the road condition using manual or automatic techniques. Manual techniques are very time consuming, so the best solution is to use innovative measuring techniques which are very efficient and accurate for detecting defects [3]. In this study mobile laser scanning was used. The dataset was acquired by the Riegl VMZ-400 scanning system. The workflow of processing the data was designed to acquire the data, post-process and analyse them to create a defect map, where cracks in the road were detected.

2 Data acquisition and data processing

To acquire the spatial data, the Riegl VMZ-400 hybrid system was used. It means that the VZ-400 scanner could be used successfully as the terrestrial and the mobile laser scanner. It helps to gather information about the entire scanned object, does not leave any space without points which is a significant assumption during planning. It is especially important in searching for deformations because all of the details of the object have to be moved to the next steps of processing the data. It is commonly known that the car cannot drive into every area so using terrestrial laser scanning as a supplement is a good solution. In circumstances, where the global accuracy should be very high, the TLS could be used as the reference [4] for mobile data.

The mobile scanner is placed on a platform. Using IMU (Inertial Measurement Unit) and a GNSS receiver the scanning system is created assisted with the DMI (Distance Measurement Indicator). To adjust all the components into one system, offsets of the device's centre have to be estimated with very high accuracy. The Riegl VMZ-400 has two mounting configurations: vertical and horizontal.

Fig. 1. Riegl VMZ-400 Vertical mounting [5].

Fig. 2. Riegl VMZ-400 Horizontal mounting [5].

To prepare a proper calibration providing correct measurements, the correlations between the devices have to be evaluated. The evaluation is based on the direction of the axis of the system (X- axis should be turned into the driving direction, Z- on the ground, Y- on the right of the platform) according to the right hand rule. The values and directions of the axis from the devices are contained in calibration papers provided by the producer. Additional measurements have to be done, according to the DMI integration with the system. With that operation, the global position of the scanner is estimated during the mission. It is crucial when the user changes the configuration frequently (e.g. vertical configuration is better for gathering information about buildings, while horizontal about roads). [6]

According to data acquisition, the values of pitch (rotation around side-to-side-axis), roll (rotation around the front-to-back-axis) and yaw (rotation around the Z axis) were established. The values of rotations are presented in Fig. 3. The accuracy of these angles depends on the proper static and dynamic alignment. The static alignment is based on 5 minutes of spatial data gathering about the position of the unit through the GNSS receiver, while the dynamic alignment consists of alternating acceleration and deceleration of the vehicle before data registration. In the project described in this paper, a proper alignment was crucial because the car was driving at 10 km/h and at that speed the estimation accuracy of the rotation values decreased. It is possible to constrain the rapidity of the drop by repeatedly executing the dynamic alignment [7], [8]. As you can see in Fig. 3, the roll angular values change from -96 to -84 degrees, pitch from 41 to 50 degrees and yaw from 50 to 350 degrees during the time between 294000 seconds to 299000 seconds. It has to be mentioned that on the right side of the picture there are charts presenting the changing values of the angles in reference to dB and Hz values.

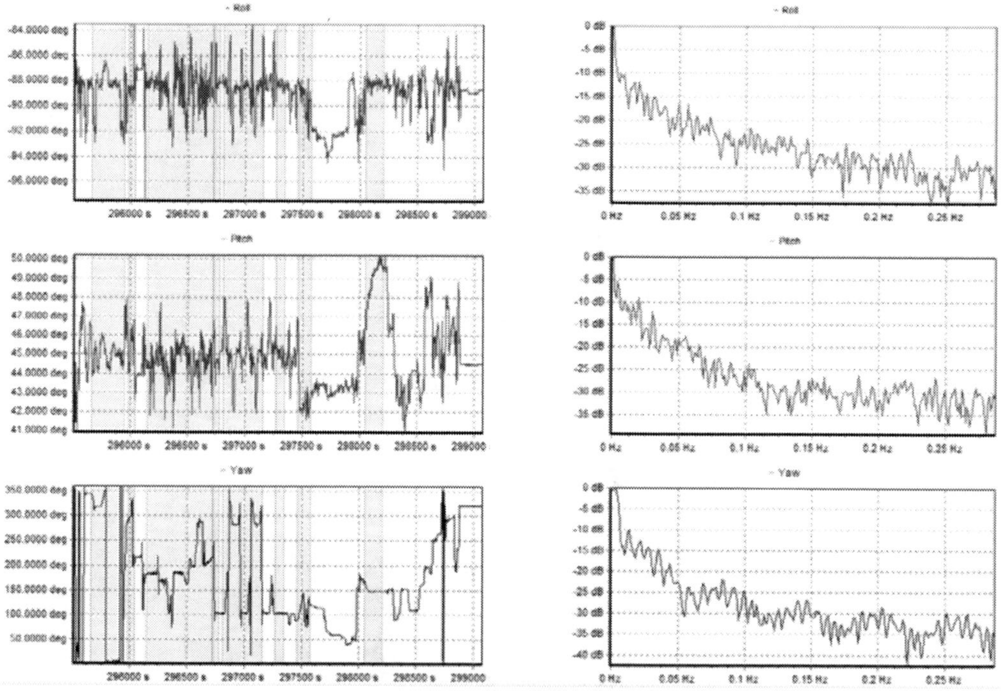

Fig. 3. Roll, Pitch, Yaw values.

The evaluation of the collected data was made after the trajectory alignment, based on GNSS reference network stations (ASG- EUPOS), but in this paper the global accuracy of the data was not that important as the accuracy of the deformations (e.g. cracks), their size and precision of the points. The result of the scanned data is presented in Fig. 4. It has to be mentioned that, when the accuracy is important, not only the reference network should be created but also the GPS positioning should be evaluated [9], [10].

Fig. 4. The result of road scanning using Riegl VMZ-400 Mobile System.

After evaluation of the data, the filtering step was made. The terrain filter used is based on a hierarchic model of data where numerous estimations were made. For each it was established if the point is located on the plane described as the ground.

The algorithm starts with a level represented by the X/Y plane of the project coordinate system and estimates robust planes in the next stages. The points which were marked during the next stages, are marked as not belonging to the ground. With that operation, it is possible to remove noise from the cars which interrupted the measurements, from vegetation if any and noise which cannot be defined.

The last preparation before searching for cracks and ruts is to triangulate the data. It is a semi-automatic process which creates surfaces by connecting the points with triangles. The results of processing operation of the point cloud are presented in Fig. 5.

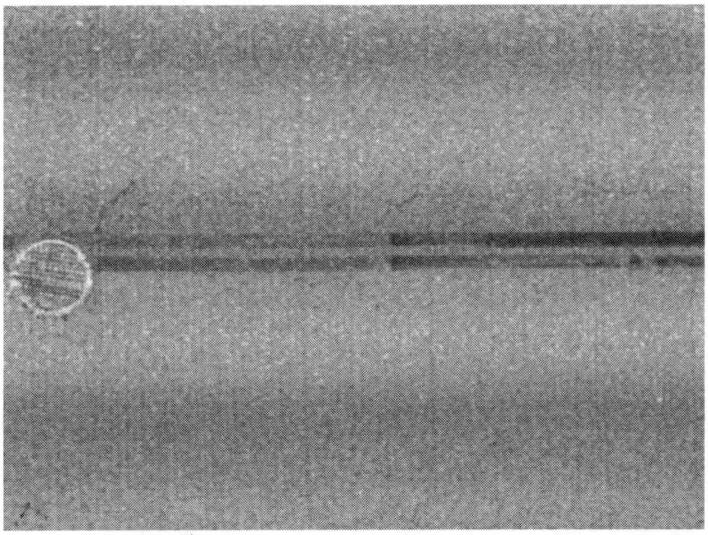

Fig. 5. The result of processed data.

3 Data analysis

The analysis of the data was based on the power of the reflected laser beam from the surface. A ray of the laser penetrated cracks on the roads because of a small diameter. The reflectance depends on the material and slope. With that property, it is possible to find cracks and ruts [13], based on material properties [14].

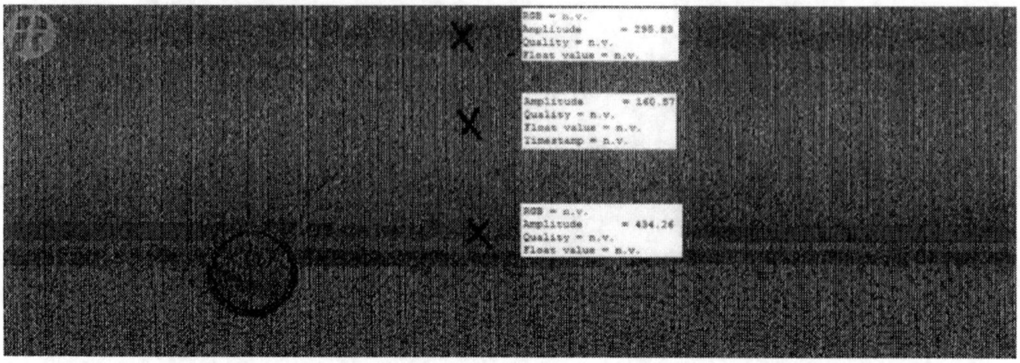

Fig. 6. Amplitude differences.

To exactly indicate places with cracks and ruts, the values of the amplitude must be estimated which are not affected by the traffic [15]. Fig. 6 shows the representation of the amplitude's values of terrain not affected, affected by the weight of traffic (based on that property the rut paths could be estimates) and cracks in the roads. What is interesting is that lines on the roads have the same amplitude as cracks. With that assumption, it is crucial to classify the road as the reflectivity function of the laser beam.

4 Experiment results

The main goal of this paper was to propose an application for searching for cracks in the road using mobile laser scanning. During the analysis of the data, the authors found, that based on the amplitude searching method there is a possibility to indicate ruts or road infrastructure such as manholes. The results of this classification are presented in Fig. 7

Fig. 6. Road classification (green - place not affected by the traffic, aqua- ruts, purple- line and cracks).

5 Conclusions

Mobile laser scanning (MLS) could be very useful for searching for road deformations. During the experiment the authors showed that the application could provide a valuable proof of the technical condition. A full analysis of the roads (for example to determine the safety factor of the paths) needs to combine geodesy measurement methods (like photogrammetry methods which are terrestrial and mobile laser scanning). Only by applying a complex and professional approach to the problem (which is estimating the roads' condition) can we indicate the technical and organizational threats and help to improve safety for all road users [13,17–19]. It has to be mentioned that if the density of the point cloud is too little to analyse, it is possible to use a high resolution point cloud collected using a different method than laser scanning [20] or by using different methods for monitoring [21].

References

1. W. Kozłowski, Innovative Technological Solutions in Terms of Realizations Partial Renovation of Bituminous Roads Using SPRIDER, Procedia Eng. **161** pp. 503–510 (2016). doi:10.1016/j.proeng.2016.08.676
2. F. Li, S.S. Liao, M. Cai, A new probability statistical model for traffic noise prediction on free flow roads and control flow roads, Transp. Res. Part D Transp. Environ. **49** pp. 313–322 (2016). doi:10.1016/j.trd.2016.10.019
3. D. Zai, Y. Guo, J. Li, H. Luo, Y. Lin, Y. Sun, P. Huang, C. Wang, 3D road surface extraction from mobile laser scanning point clouds, in: 2016 IEEE Int. Geosci. Remote Sens. Symp., pp. 1595–1598 (IEEE, 2016). doi:10.1109/IGARSS.2016.7729407
4. P. Tysiac, A. Wojtowicz, J. Szulwic, Coastal Cliffs Monitoring and Prediction of Displacements Using Terrestial Laser Scanning, in: 2016 Balt. Geod. Congr. (BGC Geomatics), pp. 61–66 (IEEE, 2016). doi:10.1109/BGC.Geomatics.2016.20
5. Riegl VMZ-400 Infosheet, www.riegl.com, (n.d.)
6. A. Kukko, H. Kaartinen, J. Hyyppä, Y. Chen, Multiplatform Mobile Laser Scanning: Usability and Performance, Sensors. **12** pp. 11712–11733 (2012). doi:10.3390/s120911712
7. A. Janowski, J. Szulwic, P. Tysiąc, A. Wojtowicz, Airborne and Mobile Laser Scanning in Measurements of Sea Cliffs on the Southern Baltic, in: 15th Int. Multidiscip. Sci. GeoConference SGEM 2015, pp. 17–24 (2015)
8. J. Szulwic, P. Burdziakowski, A. Janowski, M. Przyborski, P. Tysiąc, A. Wojtowicz, A. Kholodkov, K. Matysik, M. Matysik, Maritime Laser Scanning as the Source for Spatial Data, Polish Marit. Res. **22** (2015). doi:10.1515/pomr-2015-0064
9. A. Janowski, A. Nowak, M. Przyborski, J. Szulwic, Mobile Indicators in GIS and GPS Positioning Accuracy in Cities, Lect. Notes Comput. Sci. pp. 309–318 (2014). doi:10.1007/978-3-319-08729-0_31
10. T. Nuttens, A. De Wulf, G. Deruyter, C. Stal, H. De Backer, K. Schotte, Deformation monitoring with terrestrial laser scanning: Measurement and processing optimization through experience, in: Int. Multidiscip. Sci. GeoConference-SGEM, pp. 707–714 (Albena, 2012). doi:10.5593/SGEM2012/S07.V2006
11. J. Żukowska, Regional Observatory as a Part of Road Safety Information System, 11th Int. Conf. Transp. Syst. Telemat. (TST 2011). **239** pp. 409–416 (2011). doi:10.1007/978-3-642-24660-9_47
12. R. Bergel-Hayat, J. Żukowska, Road Safety Trends at National Level in Europe: A

Review of Time-series Analysis Performed during the Period 2000-12, Transp. Rev. **35** pp. 650–671 (2015). doi:10.1080/01441647.2015.1030005

13. A. Janowski, K. Nagrodzka-Godycka, J. Szulwic, P. Ziolkowski, Remote sensing and photogrammetry techniques in diagnostics of concrete structures, Comput. Concr. **18** pp. 405–420 (2016). doi:10.12989/cac.2016.18.3.405
14. C. Suchocki, J. Katzer, An example of harnessing Terrestrial Laser Scanner for remote sensing of saturation of chosen building materials, Constr. Build. Mater. **122** pp. 400–405 (2016). doi:10.1016/j.conbuildmat.2016.06.091
15. K. Bobkowska, A. Inglot, M. Mikusova, P. Tysi?c, Implementation of Spatial Information for Monitoring and Analysis of the Area Around the Port Using Laser Scanning Techniques, Polish Marit. Res. **24** pp. 10–15 (2017). doi:10.1515/pomr-2017-0015
16. S. Mikrut, P. Kohut, K. Pyka, R. Tokarczyk, T. Barszcz, T. Uhl, Mobile Laser Scanning Systems for Measuring the Clearance Gauge of Railways: State of Play, Testing and Outlook, Sensors. **16** pp. 683 (2016). doi:10.3390/s16050683
17. K. Jamroz, W. Kustra, M. Budzyński, J. Żukowska, Pedestrian Protection, Speed Enforcement and Road Network Structure the Key Action for Implementing Poland's Vision Zero, Transp. Res. Procedia. **14** pp. 3905–3914 (2016). doi:10.1016/j.trpro.2016.05.479
18. M. Budzyński, K. Jamroz, J. Pyrchla, W. Kustra, K. Pyrchla, Identifying the Effects of Selected Road and Roadside Parameters on Road Safety Using Geodetic Techniques, in: 2016 Balt. Geod. Congr. (BGC Geomatics), pp. 154–158 (IEEE, 2016). doi:10.1109/BGC.Geomatics.2016.35
19. B. Hejmanowska, W. Kaminski, M. Przyborski, K. Pyka, J. Pyrchla, Modern remote sensing and the challenges facing education systems in terms of its teaching, in: 7th Int. Conf. Educ. New Learn. Technol., pp. 6549–6558 (IATED, Barcelona, 2015)
20. K. Bobkowska, A. Janowski, M. Przyborski, J. Szulwic, Analysis of High Resolution Clouds of Points as a Source of Biometric Data, in: 2016 Balt. Geod. Congr. (BGC Geomatics), pp. 15–21 (IEEE, 2016). doi:10.1109/BGC.Geomatics.2016.12
21. K. Michalowska, E. Glowienka, A. Pekala, Spatial-temporal detection of changes on the southern coast of the Baltic Sea based on multitemporal aerial photographs, ISPRS - Int. Arch. Photogramm. Remote Sens. Spat. Inf. Sci. **XLI-B2** pp. 49–53 (2016). doi:10.5194/isprsarchives-XLI-B2-49-2016

AUTHOR INDEX

EDP Sciences – Web of Conferences
17, Avenue du Hoggar
Parc d'Activité de Courtabouf
BP 112
F-91944 Les Ulis Cedex A
France

ISBN 978-1-5108-4706-4

CURRAN ASSOCIATES INC.
proceedings
.com

18th International Conference Diagnostics of Machines and Vehicles (Diagnostyka 2019)

MATEC Web of Conferences Volume 302 (2019)

Bydgoszcz, Poland
12 December 2019

edp sciences